Springer Tracts in Modern Physics
Volume 167

Springer

Berlin
Heidelberg
New York
Barcelona
Hong Kong
London
Milan
Paris
Singapore
Tokyo

Physics and Astronomy

ONLINE LIBRARY

http://www.springer.de/phys/

Springer Tracts in Modern Physics

Springer Tracts in Modern Physics provides comprehensive and critical reviews of topics of current interest in physics. The following fields are emphasized: elementary particle physics, solid-state physics, complex systems, and fundamental astrophysics.

Suitable reviews of other fields can also be accepted. The editors encourage prospective authors to correspond with them in advance of submitting an article. For reviews of topics belonging to the above mentioned fields, they should address the responsible editor, otherwise the managing editor. See also http://www.springer.de/phys/books/stmp.html

Managing Editor

Gerhard Höhler

Institut für Theoretische Teilchenphysik
Universität Karlsruhe
Postfach 69 80
76128 Karlsruhe, Germany
Phone: +49 (7 21) 6 08 33 75
Fax: +49 (7 21) 37 07 26
Email: gerhard.hoehler@physik.uni-karlsruhe.de
http://www-ttp.physik.uni-karlsruhe.de/

Elementary Particle Physics, Editors

Johann H. Kühn

Institut für Theoretische Teilchenphysik
Universität Karlsruhe
Postfach 69 80
76128 Karlsruhe, Germany
Phone: +49 (7 21) 6 08 33 72
Fax: +49 (7 21) 37 07 26
Email: johann.kuehn@physik.uni-karlsruhe.de
http://www-ttp.physik.uni-karlsruhe.de/~jk

Thomas Müller

Institut für Experimentelle Kernphysik
Fakultät für Physik
Universität Karlsruhe
Postfach 69 80
76128 Karlsruhe, Germany
Phone: +49 (7 21) 6 08 35 24
Fax: +49 (7 21) 6 07 26 21
Email: thomas.muller@physik.uni-karlsruhe.de
http://www-ekp.physik.uni-karlsruhe.de

Fundamental Astrophysics, Editor

Joachim Trümper

Max-Planck-Institut für Extraterrestrische Physik
Postfach 16 03
85740 Garching, Germany
Phone: +49 (89) 32 99 35 59
Fax: +49 (89) 32 99 35 69
Email: jtrumper@mpe-garching.mpg.de
http://www.mpe-garching.mpg.de/index.html

Solid-State Physics, Editors

Andrei Ruckenstein
Editor for The Americas

Department of Physics and Astronomy
Rutgers, The State University of New Jersey
136 Frelinghuysen Road
Piscataway, NJ 08854-8019, USA
Phone: +1 (732) 445 43 29
Fax: +1 (732) 445-43 43
Email: andreir@physics.rutgers.edu
http://www.physics.rutgers.edu/people/pips/
Ruckenstein.html

Peter Wölfle

Institut für Theorie der Kondensierten Materie
Universität Karlsruhe
Postfach 69 80
76128 Karlsruhe, Germany
Phone: +49 (7 21) 6 08 35 90
Fax: +49 (7 21) 69 81 50
Email: woelfle@tkm.physik.uni-karlsruhe.de
http://www-tkm.physik.uni-karlsruhe.de

Complex Systems, Editor

Frank Steiner

Abteilung Theoretische Physik
Universität Ulm
Albert-Einstein-Allee 11
89069 Ulm, Germany
Phone: +49 (7 31) 5 02 29 10
Fax: +49 (7 31) 5 02 29 24
Email: steiner@physik.uni-ulm.de
http://www.physik.uni-ulm.de/theo/theophys.html

Rolf Könenkamp

Photoelectric Properties and Applications of Low-Mobility Semiconductors

With 57 Figures

 Springer

Dr. Rolf Könenkamp

Hahn-Meitner Institute
Glienicker Strasse 100
14109 Berlin, Germany
E-mail: koenenkamp@hmi.de

Library of Congress Cataloging-in-Publication Data.

Könenkamp, Rolf, 1954- . Photoelectric properties and applications of low-mobility semiconductors/Rolf Könenkamp. p.cm.– (Springer tracts in modern physics, ISSN 0081-3869; v. 167). Includes bibliographical references
and index. ISBN 3-540-66699-0 (hc.: alk. paper). 1. Semiconductors-Electric properties. 2. Electron transport. 3.
Photoelectricity. I. Title. II. Springer tracts in modern physics; 167. QCI.S797 vol. 167 [QC611.6.E45] 537.6'226–dc21
99-045236

Physics and Astronomy Classification Scheme (PACS): 72.20.-i, 72.40.+w, 72.80.+r,
73.20.-r, 73.40.-c

ISSN 0081-3869
ISBN 3-540-66699-0 Springer-Verlag Berlin Heidelberg New York

Springer-Verlag is a company in the BertelsmannSpringer publishing group.
© Springer-Verlag Berlin Heidelberg 2000
Printed in Germany

Typesetting: Data conversion by EDV-Beratung F. Herweg, Hirschberg
Cover design: *design & production* GmbH, Heidelberg

Printed on acid-free paper   SPIN: 10748147   56/3144/tr   5 4 3 2 1 0

Preface

This volume discusses the photoelectric behavior of three semiconducting thin film materials – hydrogenated amorphous silicon (a-Si:H), nano-porous titanium dioxide, and the fullerene C_{60}. Despite the fundamental structural differences between these materials, their electronic properties are – at least on the phenomenological level – surprisingly similar, since all three materials have rather low carrier mobilities.

In the last decade a-Si:H has conquered large market segments in photovoltaics, flat panel displays and detector applications. It is surely the most advanced and best understood of the three materials. Nano-porous TiO_2 is used successfully in a novel solar cell featuring an organic dye absorber. This product is now at the brink of commercialization, while electronic applications for C_{60} still appear to be in the exploration phase.

At this stage it appears that some of the insight and many of the experimental techniques used in the development of a-Si:H may prove useful in the on-going and yet very basic study of TiO_2 and C_{60} thin films. This idea is the guideline to this book. Without being comprehensive on the part of amorphous silicon, it attempts to outline basic characterization schemes for the nano-porous and fullerene materials, and to evaluate their potential for applications with respect to a reference, which is given by a-Si:H.

It is a pleasure to thank the many colleagues, students and friends, who took part in this research work and accompanied me over the last years. I am grateful to E. Wild, R. Henninger, R. Engelhardt, G. Priebe, A. Wahi, K. Boedecker, K. Ernst, S. Siebentritt, who worked in my group at HMI, and to C. H. Fischer, M. Giersig, P. Hoyer, B. Pietzak, A. Weidinger, and M. C. Lux-Steiner at HMI, S. Wagner and H. Gleskova at Princeton University, T. Shimada and S. Muramatsu at the Hitachi Central Research Labs in Tokyo, and C. Levy-Clement at CNRS in Paris.

Berlin, April 2000   Rolf Könenkamp

Contents

1. Introduction

Semiconductors have become the preferred material for processing, transfer and – to some extent – the storage of information. Whereas the processing is performed mostly in purely electronic devices based on mono-crystalline Si, information transfer uses electro-optical processes in III–V and II–VI compound semiconductors. The general trends in this area of technological development are cost reduction by integration and higher speed in the handling of ever larger amounts of information. These trends drive both the growth of larger crystals and the miniaturization of devices.

Perhaps less impressive, but also experiencing steady growth, is a different field of semiconductor applications that is geared toward developing large-area devices such as displays, scanners, solar cells and large-area detectors. The materials involved here are glassy, amorphous, nano-crystalline and organic semiconductors. These have in common a high degree of freedom in structural arrangement and can therefore easily be prepared as extended, isotropic and highly homogeneous thin films on a nearly arbitrary substrate. There are no grain boundaries and no preferred directions in amorphous solids, and in the nano-crystalline and organic materials grain boundaries are visually not resolved. Generally, the fabrication processes are simpler and far less costly than in crystalline technologies.

The main challenge is to obtain suitable electronic properties in these materials. Solid-state theory tells us that fast and long-range carrier transport is possible in the extremely regular structure of single crystals, while dissipation of electron energy results from scattering at imperfections. With their lack of long-range order, grain boundaries, and lattice distortions, large-area materials clearly appear to be "ideal scatterers". It is therefore not surprising that the carrier mobilities are extremely low. In single crystals, electron mobilities range up to 10^6 cm^2(Vs)$^{-1}$, and ballistic carrier ranges span thousands of lattice constants. In amorphous, nano-crystalline and organic thin films the mobilities are often far below 1 cm^2(Vs)$^{-1}$ and scattering occurs within few interatomic distances. The demand for improvement is therefore a serious one and includes both material optimization and new concepts for the design of devices.

1.1 Purpose of This Book

This book deals with just this challenge. Its aims are to clarify the basic transport schemes in low-mobility semiconductors, to determine the values of the transport parameters, to illustrate experimental methods, and to develop new ideas for photoelectric applications. The study focuses on three materials from different material classes. Hydrogenated amorphous Si (a-Si:H) is the most advanced of the three, and its electronic properties have been summarized in a number of reviews [1–3]. The other two, porous nanocrystalline oxides and the fullerenes, are still in a stage of basic research and are presently under intense investigation. The following sections contain a very brief introduction to the on-going work. Chapter 2 is concerned with the basic concepts in low-mobility transport. Electronic localization is discussed since it is of strong relevance in all three materials. It will be pointed out that a small set of parameters can serve as a form of evaluation in material assessment, and it will be shown how these parameters are determined in time-resolved and steady-state experiments.

The main part of this book will focus on the electronic transport properties of thin films, with particular reference to devices such as solar cells, thin-film transistors and charge-storage diodes. Unipolar transport concepts will be discussed in the context of thin-film transistors and charge-storage diodes, while solar cells are used to illustrate bipolar transport. The picture developed in Chap. 3 for a-Si:H will be taken as the basis for an interpretation of newer transport results for porous TiO_2 and the fullerenes. These materials are presented in Chaps. 4 and 5, which comprise a discussion of the transport properties and processes in these novel materials. There will also be some reference to recent efforts in device fabrication. For both materials the examples chosen will be solar cells [4,5].

1.2 Basic Material Characterization and Preparation

1.2.1 Hydrogenated Amorphous Silicon

In the description of amorphous materials, it is useful to distinguish between their short- and long-range properties. Amorphous Si exhibits many of the local properties of its crystalline counterpart. The coordination is also tetrahedral, with only minor deviations in nearest-neighbor distance and bond angle. The first and second coordination shells are nearly identical with those of the ideal diamond lattice. However, a well-defined third-neighbor coordination is not found [6], because the dihedral angle between neighboring tetraheders varies strongly in the amorphous case. A fully connected amorphous structure, with variations only in the dihedral angle, can be constructed when the average coordination number remains below ∼2.5 [7]. Si with a valence of

4 can therefore only form amorphous networks with a large density of co-ordination defects, such as dangling bonds and voids [8]. These defects are detrimental to the electronic properties, because they give rise to localized states at energies near the center of the bandgap [9,10]. At these energies they constitute very efficient recombination and trapping centers. Early efforts in the technological development of the material have therefore focused on passivating dangling Si bonds. This has been achieved by incorporation of H [11–13], and to some limited extent with F [14]. In optimized a-Si:H the density of dangling Si bonds is as low as $\sim 10^{15}$ cm^{-3}. The H content for such material typically lies in the range of 2–15 at.%. This material is easily doped [15–17] and can be prepared in a number of similar chemical vapour deposition (CVD) processes [18–21]. Today practically all useful devices are made from hydrogenated CVD grown material.

To study the structure and short-range order in a Si:H, Raman scattering has been successfully applied [22]. Due to the absence of phonon wavevector conservation, the Raman spectrum is noticeably different from that of crystalline Si. This allows a quantitative determination of the amorphous and crystalline content when the films are not purely amorphous, as illustrated in Fig. 1.1.

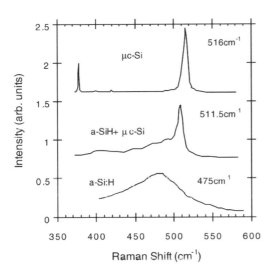

Fig. 1.1. Raman spectra of amorphous and partially crystallized a Si:H films [23]

The distribution of localized states in the band-gap of a-Si:H can be determined by deep-level transient spectroscopy (DLTS) and photoemission spectroscopy. Experiments using these methods have allowed to establish correlation between structural properties and the density of states (DOS). The DOS for n-type a-Si:H obtained from DLTS [24] is shown in Fig. 1.2. The DOS is continuous across the bandgap, with a minimum near the Fermi-energy. The integrated midgap density, approximately equals the spin density

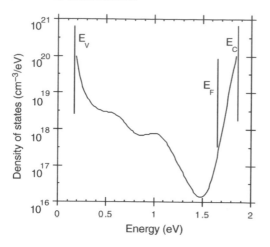

Fig. 1.2. DOS distribution in the mobility gap of n-type a-Si:H

with a g value of 2.0055 [25]. This g value is characteristic for threefold coordinated Si, indicating that the localized mid-gap states are due to dangling Si bonds. Towards the valence and conduction bands, the DOS increases exponentially, giving rise to the conduction and valence band-tails. The band-tails are associated with strained regions in the network caused by static distortions and thermal disorder [26,27]. The band-tail states are localized, but due to their high density they participate in room-temperature conduction, either by hopping among the localized states or by a multiple-trapping transport mode [1,28,29] which involves carrier emission to the band. In optimized material the valence band-tail has a higher number of states than the conduction band-tail. The energy at which the transition between localization and extended behavior occurs is called the mobility edge. Experimental data suggest that the mobility gap coincides closely with the optical [1] gap. The distribution of states in the gap appears to depend in a systematic way on the position of the Fermi-level, E_F [30]. DLTS and photoemission data on differently doped material show that the DOS minimum occurs near E_F, while a broad defect band is located near the band-edge opposite E_F. This behavior can be explained by structural relaxation of the amorphous network, driven by the trend to minimize the total energy [31]. Structural relaxation is also observed when the material is subject to photon, electron or particle irradiation [32]. In many applications the material is therefore not stable [3], and today much of the research aims at understanding and controlling the structural metastability. Other R&D topics are concerned with the optimization of alloys [33] and the development of electronic devices [3].

For an assessment of the transport properties, it is important to know how the different groups of defect states affect the carrier propagation, and how this bears on device performance.

Preparation. The a-Si:H and a-Si:H-alloy samples described in this work were prepared by plasma-assisted chemical vapor deposition of silane, germane and methane [34–37]. The preparation conditions were chosen to produce films with optimized photoconductivity. Typically, the deposition occurred at a substrate temperature of 270°C, at a pressure of 1 mbar, with 400 mW rf power between ∼10 × 10 cm electrodes, and at gas flow rates around 50 sccm. The devices were prepared in multi-chamber reactors which allowed the deposition of films with negligible dopant carry-over from layer to layer.

1.2.2 Nano-Crystalline Films

The transition from amorphous to nano- and micro-crystalline films is not sharp, and the optical, electrical and structural properties of thin films can be adjusted between the purely amorphous and single-crystalline extremes. This is particularly true for the sintered materials used here and is mainly because of two reasons: In sintered materials the grain-boundary volume usually constitutes a large fraction of the total volume, and structural distortions therefore have a noticeable influence on most film properties. Secondly, there may be amorphous regions between the crystallites, such that one actually deals with a mixed solid made up of crystalline and amorphous fractions. The ratio between the crystalline and amorphous volumes depends on growth, annealing, grain size and film thickness. With increasing crystallinity, one observes the sharpening of spectral features, in particular at the band-edges, the appearance of crystal singularities in the bands and a decreasing absorption in the band-gap. This is illustrated in Fig. 1.3, which shows a transmission electron micrograph of two grains of a TiO_2 film as used in this work. The micrograph clearly shows that the grains of these films are monocrystalline. Using Fourier transform techniques, a lattice plane separation of 2.3 Å was

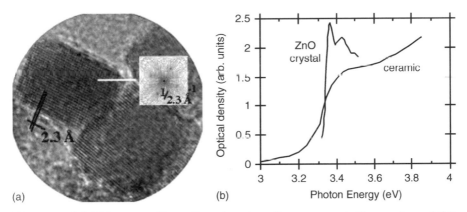

(a)   (b)

Fig. 1.3. (a) TiO_2 crystallites after sintering: the monocrystalline grains exhibit lattice distortions and roughening on the grain boundary and interfaces. The insert shows the electron diffraction pattern obtained from a Fourier transform analysis. Micrograph from M. Giersig, Hahn–Meitner Institut Berlin [38]. **(b)** Absorption edges for crystalline and nano-crystalline ZnO [39]

determined, which corresponds to the spacing of (004) planes in the TiO_2 anatas structure. Close inspection also reveals lattice distortions on the grain boundaries between the two crystallites. Figure 1.3b shows a comparison of the optical absorption edges of a ZnO crystal and a sintered nano-crystalline ZnO film [39], illustrating the broadening of optical features.

Porous Nano-Crystalline Metal Oxide Films. Many metal oxides exhibit a high transparency in the visible spectral region, and have a high dopability. In the form of compact, poly-crystalline thin films, these materials are used as transparent contacts in many photoelectric applications. One finds them as window layers in solar cells, as transparent contact lines in displays and scanners, and as area contacts to liquid crystals. Porous, nano-crystalline films have so far mostly attracted interest in photo-electrochemistry and photocatalysis. Due to the wide band gap, excited carriers provide sizable quantum energies which can be exploited in chemical reactions ranging from the oxidation of hydrocarbons [40] to electrochromic devices [41,42]. There is also interest in using porous films as templates and electron conductors for novel semiconductor hetero-junctions [43,44]. For this purpose thin films are prepared in sol-gel processes as loosely connected ceramics of nano-meter-sized crystallites. The connectivity within the film, the crystallite size, and the internal surface area can be widely adjusted by controlling the process conditions. By depositing other semiconductor materials on the internal surface of the films, hetero-junctions of extremely large area can be prepared. Such large-area junctions can also be prepared in zeolytes [45], polymers [46], inverse micelles [47] and glasses [48], but the electronic properties of the metal oxides make these particularly interesting for photoelectric applications. Large-area hetero-junctions could strongly improve efficiency in the separation of photogenerated carriers, and the distributed topology of the junction could allow for extremely short transport paths in the void-filling material. This can be a particular advantage for materials with poor electronic transport properties. The small-scale porosity allows the preparation of stable quantum dots [39,44,49–54] inside the films, which may open the way for large-area applications of confined electron systems [55]. Conventionally, quantum structures are prepared in a combination of epitaxial and lithographic techniques, since these allow a high degree of control over the growth process and the resulting device structure. So far, however, these techniques have remained restricted to wafer-size dimensions. The preparation techniques described in this book are considerably simpler and better suited for fabrication on large substrates. The main draw back, at present, is the size dispersion and the irregular distribution of the prepared quantum dots. Nonetheless, very strong confinement effects can be obtained and the thin film structures described here may therefore hold promise for a number of photoelectric applications [41,42,56].

Preparation. The porous TiO_2 films are prepared in sol-gel processes using colloidal suspensions. The TiO_2 colloid is prepared from titanium isopropoxyde and isopropanol in water. For the peptization, nitric acid is added; subsequent boiling induces the crystallization of TiO_2 particles and the evaporation of organic components [39,57,58]. Crystallites with an average diameter as low as ∼10 Å can be prepared in this way. Crystallites with size below 1 μm are mono-crystalline and have a fairly rounded shape.

The colloid is the feedstock material for the preparation of thin films. Simple deposition techniques, such as spin coating and screen printing are used to prepare colloidal layers on conducting or insulating substrates [39]. The deposited films are subsequently heated for several minutes at temperatures around 450°C. Deposition and drying are repeated until the desired thickness is obtained. Subsequent sintering at temperatures up to 650°C results in porous ceramic-type films consisting of 40–80 Å diameter anatas crystallites. Surface adsorption experiments show that the inner surface of a 4 μm thick film can be 1000 times larger than the projected surface area.

Coating of these porous templates with compound semiconductors is carried out in gas-phase CVD and in electrochemical or purely chemical methods involving liquid solutions. A comparison between gas- and liquid-phase deposition indicates that only deposition from the liquid phase produces an intimate large-area contact between the nano-porous substrate and the newly deposited layer. Gas-phase deposition leads to a nearly planar super-structure without any significant material penetration into the substrate. Electrodeposition enables mass transport deep into porous films when the void size is ∼100 Å or larger. When pulsed deposition techniques with duty cycles around 1% are applied, the penetration into the films is found to be further improved. In films with smaller void size, electrodeposition becomes increasingly less effective.

The best conformal coverage is obtained in bath depositions, even for voids as small as 50 Å. Typical processing procedures for compound semiconductors involve the deposition of a metal-ion layer from a suitable metal-salt solution and the subsequent chemical reaction to form the semiconductor. Ideally, the metal is deposited in its final oxidation state to improve the reaction kinetics and prevent the formation of undesired by-products of other stoichiometry. This preparation scheme is suitable for the deposition of connected films [59,60] and quantum dots [61–63]. II–VI semiconductors such as CdTe, CdS and CuS_2 have been prepared as connected thin films, and PbS as clusters in the quantum dot regime. For quantum dots, the deposition of the metallic precursors and the subsequent compound formation step have to occur at sufficiently low rates in order to prevent the formation of a continuous layer. PbS quantum dots had a minimum size of ∼30 Å, but larger quantum dots can be prepared by agglomeration in repeated deposition and heating sequences [44,61].

1.2.3 Fullerenes

The discovery of C_{60} was one of the major surprises of recent times in physics and chemistry [64,65]. It was quite unexpected to find yet another stable, highly symmetric and naturally occurring form of carbon after the many centuries of work on this element. Beyond the highly symmetric C_{60} molecule there exist a number of related molecules of higher weight and lower symmetry, such as C_{70}, C_{76}, C_{78}, C_{84}, C_{90} etc. [66]. Beyond the single-walled species there are also structures with double or more-fold walls, and tube-like structures with closed or open ends (nano-tubes) [67,68]. Research has focused on structural modifications such as endohedral intercalation [69,70] and chemical complexation [71,72]. A broad range of applications have been suggested, including structural and tribologic applications [73], the medical use of C_{60} [74], diamond formation [75,76] and initial semiconductor applications [77,78]. As a solid, C_{60} forms an isotropic 3-dimensional molecular crystal with simple-cubic structure below 249 K and face-centered cubic structure above [79]. Neutral C_{60} without additional impurities is a semiconductor, but when doped properly it can be made metallic [80] or superconducting [81]. In the solid, the molecule exhibits 6 oxidation states [82]; organic compounds with C_{60} in the oxidation state -2 exhibit weak ferromagnetism [83].

The key to the description of the electronic properties of films is the fullerene molecule, since the inter-molecular wavefunction overlap in the solid is fairly small. Any attempt to characterize the electron transport in the solid therefore has to include a description of the molecular excitation scheme in addition to the lattice properties. It will be seen that the excitation spectra of the solid bear close resemblance to those of the isolated molecule. Solid-state properties are therefore best described in the excitonic framework rather than using the independent electron approximation. The excitonic character opposes separation of excited electron–hole pairs and gives rise to fast carrier recombination and localization. It appears that efficient electronic transfer is only obtainable in hetero-junction structures [84].

Preparation. Thin-film preparation involves the sublimation of C_{60} powder, which is typically achieved in a low-pressure, carbon-arc process [85]. The powder is chromatographically purified and deposited by sublimation at 550°C in ultra-high-vacuum (UHV) conditions with the substrates kept at room temperature. This procedure gives polycrystalline films containing >99.9% C_{60} [86] with grain size between 70 and 200 nm. An atomic force micrograph depicting the grain structure of a 100-Å-thick film is shown in Fig. 1.4.

For experimental reasons many of the measurements employed in this work had to be carried out outside the UHV chamber. Since oxygen has a strong effect on the electronic properties of C_{60}, the films needed to be covered with an oxygen-diffusion barrier. In the case of sandwich structures, a 100-Å metal film, deposited without breaking the UHV conditions, served

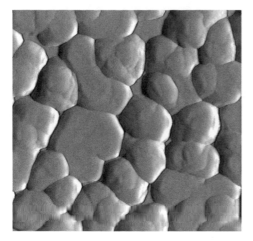

Fig. 1.4. Atomic-force micrograph of a 100-Å-thick C_{60} film grown on glass substrate. The field of view has a width of 1 μm

this purpose. Films with coplanar contacts were sealed with polymer tape inside the load-lock chamber at pressures of $\sim 10^{-5}$ mbar [87,88].

2. Basic Concepts of Low-Mobility Transport

2.1 Localization

Since its first discussion by Anderson [89,90], the question of electron localization has been approached from a number of different perspectives, such
as percolation theory, scaling theory and field theory [91–94]. Here a brief
sketch of the basic results of scaling theory as proposed by Abrahams et al.
[95,96] is given. This approach starts out by developing the concept of microscopic conductance, and in a second step extends this idea to macroscopic
dimensions.

In macroscopic dimensions one expects conductance to scale linearly with
scattering time, τ, as given by the Drude expression for the conductivity,

$$\sigma = ne\mu = ne^2\tau/m \tag{2.1}$$

where n is the carrier density, e the elementary charge, μ the mobility, and
m the effective carrier mass. As τ approaches zero, (2.1) gives a linearly
vanishing conductivity. The uncertainty principle, however, predicts that the
carrier energy changes as

$$\Delta E > \hbar/\Delta t \tag{2.2}$$

when the time scale decreases. This increase in energy allows the carrier
to access states which differ from its own energy, thereby opening a new
quantum-mechanical route for carrier transfer. Since only states within an
energy interval ΔE are accessible, ΔE must be greater than the average
level spacing for conduction to take place. Denoting the level spacing as W,
the criterion for transport is $\Delta E \gg W$, and for localization $\Delta E \ll W$. To
make this argument more quantitative, let us consider that the transport is
non-directional and diffusive. The sampled volume is then given by L^3 with L
being the diffusion length, which depends on the diffusivity, D, and on time
as

$$L = (D\Delta t)^{1/2}. \tag{2.3}$$

When $N(E)$ is the DOS, the average level spacing in the sampled volume is
given by

$$W = (N(E)L^3)^{-1}. \tag{2.4}$$

By use of the Einstein relation, the mobility can be expressed as $\mu = eD/kT$. The conductivity is then given by $\sigma = ne\mu = e^2nD/kT$ or, with (2.2) and (2.3),

$$\sigma = \left(\frac{e^2}{\hbar}\right) nL^2 \frac{\Delta E}{kT}. \tag{2.5}$$

When the transport occurs at the Fermi-level, the carrier density is $n = N(E_F)kT$. Inserting this and (2.4) into (2.5), one obtains for the conductance

$$g = L\sigma = \left(\frac{e^2}{\hbar}\right) \frac{\Delta E}{W(L)}, \tag{2.6}$$

which confirms the interpretation of $\Delta E/W$ and suggests $e^2/\hbar$ as the natural units for the conductance. Equation (2.6) gives an expression for the conductance in microscopic volumes of size L. The conductance depends on the density of states in the accessible volume. If the DOS is a function of energy, one may expect a transition between localization and conductive behavior at a critical energy. This energy is the mobility edge. The scaling behavior of the conductance, i.e., the dependence of g on the sample size, can be determined by considering general physical requirements for the scaling function and knowledge about the conductance for macroscopically large samples [95].

When the scattering is very strong and the carrier energy is lost within one scattering event, ΔE in (2.1) is given by $\Delta E = \hbar/\tau$ with τ being the scattering time rather than the diffusion time. Accordingly, L in (2.5) and (2.6) has to be replaced by the scattering length. As the smallest possible scattering length is given by the nearest-neighbor distance, it follows that the nearest-neighbor distance sets the lower limit for the scaling behavior [97,98].

Localization Due to Disorder. The influence of disorder on electron diffusion properties was first treated by Anderson [89], by studying effects of a random potential superimposed on a periodic potential. It was shown that electron diffusion vanishes (at zero temperature) when the random fluctuations, ΔV, are larger by a factor of 3 than the size of periodic potential, i.e., $\Delta V > 3V_L$. Since the bandwidth broadening is approximately given by $(V_L^2 + \Delta V^2)^{1/2}$, localization in disordered solids sets in when their bandwidth is ~3 times that of the corresponding crystal. Since the total number of states is constant, there is a peak density of states in the band which is ~1/3 that in the crystalline case. Experimental results on amorphous materials show, however, that typical potential fluctuations are only of the order of ~0.5 eV, while typical bandwidths in inorganic materials are >5 eV. Thus, the condition for complete localization is not met in these materials. However, disorder still induces a band broadening that becomes evident in the form of band-tails. When the DOS in the band-tails falls below the critical value for localization, the corresponding states are localized. Weak disorder thus results in localization for a fraction of the bands, and the mobility edge separating these from the extended states is located in a continuous density of states. In Si, where the DOS in the band is 2×10^{22} cm^{-3} eV^{-1}, the Anderson criterion predicts the mobility edge to occur at $\sim 7 \times 10^{21}$ cm^{-3} eV^{-1}.

Transport in Localized States. There are essentially two ways in which localized states can contribute to the carrier transport at finite temperatures. Since the band-tail states are so close to the extended states in the band, transport can result from thermal emission from the localized tail states. Transport then occurs in extended states until recapture in localized states occurs. This process occurs in a repeated fashion, the kinetics being mainly determined by the thermal emission rates. These rates depend exponentially on the energy difference between the state and the band, E, as

$$\nu(E,T) = \nu_0 \exp(-E/kT), \tag{2.7}$$

where ν_0 is an attempt to escape frequency in the phonon range. This transport mode is known as multiple trapping.

There is also a sizable probability for direct transitions among localized states. The corresponding mode of propagation is known as hopping conduction. In Mott's model of hopping transport [28] the transfer probability between localized sites separated by distance ℓ and energy ΔE is given by,

$$p = \nu_0 \exp(-2\ell/\ell_0 - \Delta E/kT), \quad \text{for } \Delta E > 0,$$
$$p = \nu_0 \exp(-2\ell/\ell_0), \quad \text{for } \Delta E < 0, \tag{2.8}$$

where ℓ is the distance between localized sites, and ℓ_0 is the extension of the wavefunction [28,92–94].

Which of the two transport modes, multiple trapping or hopping, dominates will depend on the temperature and the details of the DOS distribution. In some cases both transport modes give similar contributions, as was pointed out by Monroc [99]. Due to the exponential factors in (2.7) and (2.8), transport involving localized states is characterized by an extremely broad distribution of time constants, which results in unusually large broadening of initially narrow carrier distributions. Transport under such conditions is called dispersive.

It appears reasonable to apply similar ideas to the transport in the nanocrystalline materials, since one finds bandtails in practically all optical spectra of these materials [100]. To what extent the band-tails have an influence on the carrier propagation depends largely on their depth, however. If the bandtail depth is comparable to or larger than the ambient thermal energy, one can expect strong effects on transport, since a large portion of thermalized excited carriers then occupy band-tail states. As kT becomes larger than the band-tail depth, capture still occurs, but the emission rates are very high and the average occupation time of the band-tail states decreases, which fosters extended-state-like transport.

It is interesting to note that there is evidence for disorder-related band-tails even in high-quality, mono-crystalline material [101]. These affect the transport behavior only at very low temperatures, however [102].

Localization in Organic Materials. Disorder-induced localization is also of relevance in organic materials when they are prepared as nano-crystalline or amorphous samples. For these cases it is natural to expect the occurrence of band-tails around the band extrema. In most cases, however, is is difficult to obtain spectroscopic evidence for their existence, since the band-tail region is dominated by vibronic and excitonic structure.

Another cause for carrier localization in molecular crystals lies in the small wave-function overlap between the molecules. In many cases the lower lying states keep their molecular character, and remain localized, while only the higher excited states form a band that allows significant charge transfer. Carrier transfer is often best described in the excitonic framework, which takes account of several electron correlation effects. The most evident of these is Coulomb attraction, which keeps the excited electron–hole pair in a bound state [103]. In addition to the electronic correlation, there is in many organic materials a strong coupling between electronic and vibrational excitations. These interactions enhance localization by lowering the electron energy due to structural relaxation. The formation of polarons and charge and spin density waves is a consequence of this electron–phonon coupling [104]. Exciton and polaron formation are, of course, also important in inorganic solids and many manifestations of these are known. The binding energies are, however, usually smaller in inorganic materials and the carriers are therefore mostly not bound at room temperature.

2.2 Definition of Transport Parameters

To illustrate the salient features of the transport process and define the transport parameters, a semi-classical picture for the temporal sequence of events after photon absorption is often sufficient:

The absorption of a photon initially creates a correlated electron–hole pair whose wavefunctions overlap. Due to its close proximity the electron–hole pair is bound by the Coulomb force. When the attraction is strong, or when the coupling to other states of the solid is weak, there is a large probability that the excited pair will remain bound as an exciton. Although excitons cannot contribute directly to electric currents, they may dissociate in a second process and produce free carriers. Typical processes involved are scattering by phonons, other excitons, or free carriers and extrinsic processes such as scattering at impurities or surfaces or ionization in external fields.

The ratio of absorbed photons to generated-free-carrier pairs is the quantum generation efficiency. In inorganic solids nearly all geminate pairs diffuse apart, and the quantum efficiency is close to unity, while in organic materials the quantum efficiencies can be much smaller.

In the diffusion process following the excitation step, the hot carriers dissipate their excess energy mostly by scattering with phonons or molecular vibrations [105]. In the thin film materials described here the scattering lengths are small, typically a few atomic spacings, and carrier thermalization therefore takes place over only a very short time scale. Typically within the first pico-second a thermal distribution around the bandedge has built up [106,107]. The thermalization brings about an increasing population of localized states. Capture into these states is fast, but several capture and emission processes are necessary to establish a thermal distribution among the localized states. The rate-limiting step for the redistribution is the emission process, which depends exponentially on the depth, E, of the localized state,

$$t = \nu_0^{-1} \exp(-E/kT). \tag{2.9}$$

For $E = 100$ meV one finds $t = 5$ ns, but for $E = 200$ meV the thermalization time is already in the micro second range. For the kinetic description of thermalization, it is useful to separate between the thermalized and frozen-in carriers. The separation line is referred to as the demarcation energy; its time dependence is given by $E(t) = kT \ln(\nu t)$.

Due to their lower energy, the localized states have a large occupation probability. Their density is also much larger than typical densities of excited carriers. Accordingly, a large portion of carriers are in localized states. At finite temperatures, carrier propagation occurs by hopping or multiple-trapping transport. With progressing time there is an increase in the probability that the carrier is captured by a deep localized state, from which it is not emitted for a very long time or in which it recombines with a carrier of opposite charge. It is at this point that the transport is effectively terminated, and the term carrier range refers to the corresponding distance travelled.

Mobilities and lifetimes can be defined either as averages over the total carrier density, including free and trapped carriers, or as microscopic parameters based on free carriers. The averaged parameter for the mobility is the drift mobility, μ_d, given by the extended-state mobility, μ_0, weighted by the ratio of free carriers, n, to the total carrier density, n_tot,

$$\mu_\mathrm{d} = \mu_0 n / n_\mathrm{tot}. \tag{2.10}$$

The term lifetime, τ_0, denotes the average time for a free carrier to be captured by a localized state. In ballistic transport the lifetime is given by

$$\tau_0 = (N v \sigma)^{-1}, \tag{2.11}$$

where σ is the capture cross section, v is the thermal velocity and N is the concentration of the capture centers. Depending on the experimental situation, the capture process may be associated with recombination or trapping processes; τ_0 is then referred to as the recombination or trapping lifetime.

The dynamic response of the total carrier system is given by the response time,

$$\tau = \tau_0 n_\mathrm{tot} / n, \tag{2.12}$$

which may be significantly larger than τ_0.

The materials considered here have comparably large localized state densities near the band edges. Since the capture in these shallow states is fast, it is – in many cases – experimentally not possible to distinguish the shallow trapped carriers from carriers in extended states. Most experiments thus probe the capture kinetics of the ensemble of shallow trapped and free carriers, rather than the free-carrier kinetics. The kinetic parameters of these shallow carriers are also often termed recombination and trapping times, although they are – strictly speaking – response times. Here we will refer to the lifetimes of the free-carriers explicitly as free carrier lifetimes, and otherwise follow common practice in denoting the shallow-carrier kinetics mostly in terms of recombination and trapping times, rather than response times.

The product of mobility and lifetime is the material parameter relating to the carrier range. In purely diffusive transport the $\mu\tau$ product appears in terms of the diffusion length, $L_D = (\mu\tau kT/e)^{1/2}$, while in the case of drift in electric fields, $\mathcal{E}$, it determines the Schubweg, given by $s = \mu\tau\mathcal{E}$. It is noted that (2.10) and (2.12) indicate that drift mobility and response times may depend on the carrier densities, while the $\mu\tau$ product does not, i.e., $\mu\tau = \mu_0\tau_0$.

2.3 Experiments to Probe the Carrier Transport

A number of experiments are available for the determination of electron or hole transport in semiconductors, but not all of the techniques have sufficient sensitivity to study low-mobility materials. The high sensitivity of photoelectric methods rests on the generation of excess carriers and subsequent observation of these excess carriers in the measurement. The signal strength can therefore be increased by working at higher excitation levels. Since many applications for thin-film semiconductors are photoelectric, these methods have the additional advantage that the experimental parameters can be adjusted to match realistic device conditions. In many cases a combination of time-resolved and steady-state methods allows a nearly complete determination of the transport parameters.

In time-resolved photoconductance methods, a fast laser pulse is applied to the film and the temporal changes of the induced conductance are monitored. The decay of the conductance is due to carrier capture in shallow or deep trap states, recombination, or extraction at a contact. Accordingly, capture-time constants, recombination times, or transit times can be determined. In the time-of-flight arrangement, shown in Fig. 2.1, the semiconductor film is sandwiched between two (transparent) contacts, an electric field is applied across the film, and strongly absorbed light is used to generate a narrow carrier distribution at the surface of the film. Depending on the polarity of the applied field either electrons or holes are driven across the sample. The experiment allows the determination of the carrier drift mobility and the $\mu\tau$ product. With some experimental modifications, the generation efficiency

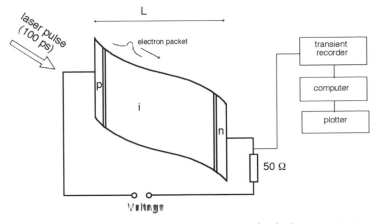

Fig. 2.1. Experimental set-up for time-resolved photoconductance and time-of-flight experiments

and the electric-field profile can also be determined. Usually the experiment is sufficiently fast that dielectric relaxation [108] can be neglected, i.e., there is no screening among the excited or thermal carriers. These conditions have the advantage of being easily evaluated, since the applied field simply super-imposes on the internal field, and the transport parameters of the two carrier types are not coupled. The dielectric relaxation time is given by $t_R = \varepsilon\varepsilon_0/\sigma$. With a conductivity, $\sigma = 10^{-7}$ $(\Omega\,\mathrm{cm})^{-1}$, and a dielectric constant, $\varepsilon = 10$, the relaxation takes approximately 10 μs.

Steady-state photoconductance experiments are often performed in the opposite limit, namely that complete dielectric and ohmic relaxation occur. In the case of spatially varying carrier densities, screening by thermal and excited carriers must therefore be accounted for, and transport and recombination of the two carrier types can no longer be considered independently. Transport under such conditions is termed ambi-polar. Depending on the experimental conditions, such as contact configuration, applied electric fields, etc., the transport behavior may be governed by either the majority or the minority carriers. In these limiting cases an evaluation of the ambi-polar transport in terms of the electron and hole parameters may be fairly straight-forward. In most other cases, however, the ambi-polar parameters become somewhat complicated expressions of the individual carrier parameters.

In this study several steady-state experiments are used. Spectrally re-solved steady-state photoconductance is employed to determine the energy levels in C_{60} films. These experiments probe majority carrier transport and are performed using the lock-in technique with chopped excitation, adjusted such that steady-state conditions are established within a single excitation cycle.

Carrier mobilities and recombination times in a-Si:H and C_{60} are determined by the moving-photocarrier-grating technique (MPG) [109,110]. This method allows the hole and electron drift mobilities and the sum of the $\mu\tau$ products to be determined. Since MPG is a novel experimental method, it is briefly described here. The experiment is shown in Fig. 2.2 and involves two coherent laser beams. The two beams have different intensities and are separately shifted in frequency by acousto-optical modulators. The beams interfere on the sample surface, where they produce an interference grating that moves at constant velocity across the sample. The grating velocity is given by $v = \lambda \Delta f / \sin \delta$, with δ being the angle between the two beams, Δf their frequency difference, and λ the laser wavelength. Due to the different intensities, the grating is superimposed on a constant intensity background.

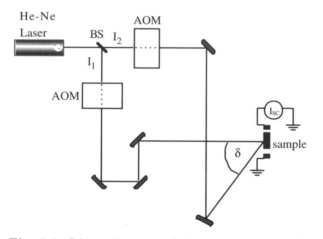

Fig. 2.2. Schematic set-up of the moving-photocarrier-grating technique (AOM: acousto-optical modulator, BS: beam splitter, I_{sc}: short-circuit current)

The moving grating is used for the photo-excitation of excess electrons and holes in the semiconductor sample. Due to their different mobilities, electrons and holes respond to the moving grating at different rates; this results in a net motion of charge, which is picked up as a photocurrent between two coplanar contacts. The dependence of the photocurrent on grating velocity is evaluated in terms of the drift mobilities of the two carrier types and the ambipolar recombination time by solving the transport and Poisson equations. The procedure is described in detail in [110].

A third type of quasi-steady-state experiment, is the junction-recovery method [111–114], which is also briefly outlined here. In the experiment both carrier types are injected through the contacts of a diode in forward bias, and then extracted by switching the diode into the reverse regime. From the kinetics of this process and the amount of stored charge, one is able to obtain information about recombination times and $\mu\tau$-products.

The forward current, I_F, is the steady-state injection current of holes and electrons through the two contacts. If the contacts block minority carrier transfer, charge neutrality implies that the injected hole and electron currents are equal. Recombination is the only loss mechanism in the device, and the stored charge is given by $Q = I_F/\tau$, where τ is the ambi-polar recombination time of the shallow carriers. The stored charge is determined by integrating the excess current transient as the diode is switched from forward to reverse. An analysis of the extraction process also allows the $\mu\tau$ product to be determined [113,114]. Figure 2.3 shows typical charge-recovery transients obtained in a-Si:H diodes.

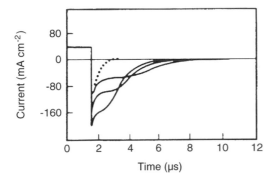

Fig. 2.3. Junction-recovery transients obtained for an a-Si:H diode after switching from a forward bias ($50\,\mathrm{mA\,cm^{-2}}$) to different reverse biases. It is noted that the recovered charge depends on the reverse voltage. This dependence is exploited to determine the $\mu\tau$ product for recombination

A determination of transport parameters under recombination conditions is also possible in the surface-photovoltage method, which yields the ambipolar diffusion length [115–118]. The method uses the photovoltaic effect at a semiconductor–electrolyte interface. Minority carriers, generated in the sample bulk, diffuse to the interface and thereby generate a photocurrent. By changing the wavelength of the generating light, the absorption depth is varied and the diffusion length can be determined from the efficiency of minority transport to the surface.

3. Transport in a-Si:H and Its Alloys

3.1 Basic Experimental Results

Basic information on the carrier transport in a-Si:H is obtained from the temperature dependence of the conductivity. In Fig. 3.1 we show an Arrhenius plot of the dark and photoconductivity in undoped a-Si:H. In the region 250 K $< T < 350$ K, the dark conductivity shows activated behavior, given by $\sigma = \sigma_0 \exp(-E_a/kT)$. The activation energy, E_a, can be taken as an approximate measure of the energy difference between the Fermi level, E_F, and the band edge, E_c. An accurate determination of $E_c - E_F$ is more complicated, since the Fermi energy, the mobility edge and the band-gap are also temperature-dependent and contribute to the conductance–temperature dependence [2,119]. Further complications arise from the fact that the activation process involves a broadened distribution of states [120–122] and that multiphonon processes are involved in the transitions [123]. In undoped material the activation energy is usually slightly less than half of the band gap, and the material is naturally n-type. In defect-rich material the activated behavior levels off towards lower temperatures, and gives rise to a T-dependence, given by $\sigma(T) = \sigma_0 \exp[(-A/T)^{1/4}]$, which is usually attributed to variable-range-hopping.

Changes in the conductivity behavior can be induced when the material is heated to temperatures above ~450 K: As Fig. 3.1b indicates, the room-temperature conductance depends on the cooling rate when the sample is heated to ~450 K and subsequently quenched. The observed behavior is particularly evident in doped material and is attributed to thermally induced structural changes [124,125], which are frozen-in at $T \approx 420$ K. The behavior is quite similar to that observed in glasses near the glass transition temperature [126]. There is now a large amount of experimental data which shows that structural changes can also be induced by illumination, carrier injection and various other treatments [32], a number of detailed models have been suggested to describe the physics underlying these changes [127–134].

Figure 3.1 shows that a-Si:H is highly photoconductive. At intensities of ~100 mW/cm^2 the photoconductance is typically 4–6 orders of magnitude larger than the dark conductance, giving typical photoconductance values in the 10^{-4} $(\Omega\,\mathrm{cm})^{-1}$ range. The temperature dependence of the photoconduc-

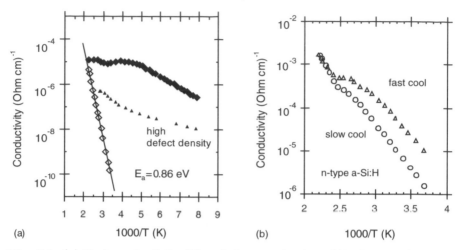

(a)   1000/T (K)   (b)   1000/T (K)

Fig. 3.1. (a) Dark conductivity (◊) and photoconductivity (♦) of undoped a-Si:H. In device-grade material the activation energy corresponds to approximately half of the energy gap. For defect-rich material (▲) the dark conductivity is given by an $\exp[(-A/T)^{1/4}]$ law, indicative of hopping transport. (b) Dark conductance of n-type a-Si:H after cooling with different rates from $T \approx 250°C$

tance can be attributed to the T-dependence of the drift mobility and the occupation of states [135].

3.2 Time-Resolved Photoconductance

Although much can be learned from a detailed evaluation of steady-state conductance, more direct insight into the transport properties and the effect of localized states comes from work on time-resolved photoconductance. Figure 3.2 shows typical results for the transient photoresponse in a time-of-flight experiment in which electrons are made to drift through an undoped a-Si:H film [136]. On a linear scale, one observes a rather featureless decay behavior; only on a log-log scale does one distinguish two time regimes. The interpretation of such transients is as follows: After thermalization of the hot, excited carriers [137,138], the photogenerated electrons occupy, predominantly localized band-tail states, only a few carriers being in extended states. In this situation the electron energy is still much higher than in thermal equilibrium, but further thermalization is kinetically slowed down due to the long thermalization times for localized states. When thermal re-excitation is the dominant process, as is the case in a-Si:H, the residence time in the traps becomes an exponential function of the trap depth. As a consequence, carriers in deep traps are essentially immobile, and only carriers in the shallow levels acquire a thermal distribution. Figure 3.3 shows the essentials of this process. Due to the broad range of time constants encountered in the thermalization

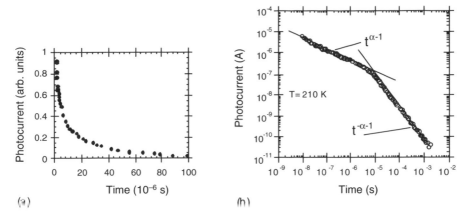

Fig. 3.2. Photocurrent transients typically observed in a-Si:H at $T = 210$ K, (**a**) Linear scale; (**b**) same transient on a log–log scale [136]

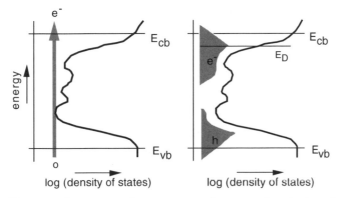

Fig. 3.3. Schematic of excitation and temporal evolution of excess carriers in the localized states between the mobility edges. Above a demarcation energy, E_D, here shown for electrons, the carriers are thermalized, while below E_D the occupation of states depends on the capture cross sections

process, the kinetics do not follow an exponential time law, but exhibit an algebraic time dependence of the kind $y(t) \sim t^{-\gamma}$ [139–141].

3.2.1 Trapping in Shallow Band-tail States

The initial decay regime in Fig. 3.2, given by the $t^{\alpha-1}$ law, corresponds to a multiple-trapping process in the shallow states of the band-tail. Here α is known as the dispersion parameter. For the case of an exponential bandtail with a bandtail parameter E_0, α is given by $\alpha = kT/E_0$. In the initial regime the total carrier density remains constant, since recombination and extraction at contacts occur only at later stages. The constant carrier density regime ends as the front portion of the carrier distribution reaches the back contact. In Fig. 3.2 this occurs after approximately 8 μs. From this time on electrons are extracted from the film, and the total number of carriers decreases. As a consequence the decay accelerates and turns into the $t^{-\alpha-1}$ decay. The rate-limiting step in the kinetics is, however, still the thermal emission from localized states, and it is therefore not surprising that the exponents in the two decay regimes are related. The correlation is not specific to carrier loss at contacts, it is encountered whenever an additional loss process, such as recombination and capture by deep states [142,143] occurs.

Algebraic decay kinetics are the signature of exponentially shaped band-tails, and these appear to be prominent in a-Si:H. The correlation between the transient shape and the DOS is, however, not a very sensitive one, i.e., any features in the DOS distribution will give rise only to small effects in the transient decay [144].

3.2.2 Drift Mobility

The time-of-flight experiment is also suited to determine the electron and hole drift mobilities. In most cases the transit time is measured as a function of applied field, E, and the drift mobility is evaluated from $\mu_d = L/(t_T E)$, with t_T being the transit time and L the sample thickness. This procedure has the advantage of being independent of the density of created electron-hole pairs, whose determination is often complicated by field-dependent quantum efficiencies, optical losses, etc.

Figure 3.4 shows typical results for holes in a-Si:H [136]. At room temperature drift mobilities in the range of 0.005 to 0.05 $\text{cm}^2(\text{Vs})^{-1}$ can be considered typical. Analysis of time-of-flight experiments shows that the transit time is usually proportional neither to the thickness of the sample nor to the inverse of the electric field. This is a direct consequence of the thermalization process within the trap-state distribution: As the average energy of the carrier distribution is lowered in the thermalization process, the drift velocity, and hence the mobility, decrease with time and are therefore time-dependent. A detailed

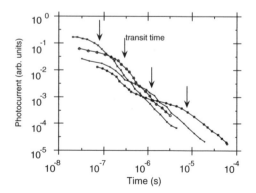

Fig. 3.4. Hole time-of-flight transients in undoped a-Si:H for different external fields at room temperature. From the transit times (*arrows*) a drift mobility of 0.012 $\text{cm}^2(\text{Vs})^{-1}$ is deduced [136]

Table 3.1. Material parameters determined from transport experiments in a-Si:H

DOS at mobility edge (E_C and E_V):	6×10^{21} cm^{-3} eV^{-1}
band gap:	1.77 eV
dangling bond density in device grade a-Si:H:	$>10^{15}$ cm^{-3}
conduction band tail parameter:	25 meV
valence band tail parameter:	50 meV
capture cross-sections:	
neutral dangling bond:	10^{-15} cm^2
tail states:	10^{-15} cm^2
charged dangling bond:	10^{-14} cm^2
free electron mobility:	15 cm^2(Vs)$^{-1}$
free hole mobility:	1 cm^2(Vs)$^{-1}$

analysis of all available experimental data has allowed a transport model to be set up that provides a quantitative determination of the DOS distribution, the capture cross-sections and the free-carrier mobility [140,145–147]. The most commonly accepted of these parameters are listed in Table 3.1.

In high-quality a-Si:H the conduction band tail parameter is only ~25 meV; electron transport therefore ceases to be dispersive at room temperature and the field and thickness dependence of the electron drift mobility approach normal behavior. As a consequence, the photocurrent transients need no longer be evaluated on a logarithmic scale, since they show a marked transition at the completion of the transit, as expected for non-dispersive transport. Non-dispersive time-of-flight transients obtained on 0.5 μm thick solar cells are shown in Fig. 3.5.

At low temperatures, $T < 100$ K, and high electric fields, $\mathcal{E} > 10^5$ V cm^{-1}, an additional field dependence of the carrier mobility has been measured which is not explained by the simple multiple-trapping model as discussed above [148–152]. Quantitative analysis of the data shows that the free-carrier mobility increases as $\mu_0 \sim \mathcal{E}^\gamma$ in undoped a-Si:H and in similar fashion in doped material [153]. The experimental results can be explained in terms of

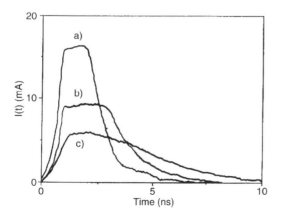

Fig. 3.5. Electron time-of-flight transients under non-dispersive transport conditions, measured in pin-type solar cells, (*a*) 3.2 V and (*b*) 1.6 V applied bias; (*c*) transient due to the built-in potential of the cell. A drift mobility of 0.46 cm^2(Vs)$^{-1}$ is deduced [136]

a field-assisted hopping process which dominates the transport behavior at low temperatures because thermally activated processes are frozen out [99]. Application of a strong electric field affects the re-emission from traps and the energy distribution of the carriers, thereby leading to an effective increase in the carrier drift mobility.

3.2.3 Electric Field Profiles

When the transport is non-dispersive, the evaluation of the transient shape also allows for a determination of the electric field profile, $\mathcal{E}(x(t))$ [154–156]. The idea of the experiment is straightforward: Since the mobility is constant in time, the time dependence of the current can be attributed solely to the non-uniform field, $\mathcal{E}(x)$, which is sampled by the drifting carriers in the course of the transit. In terms of the photocurrent, the position x of the carrier packet is given by

$$x(t) = \int v(t)\mathrm{d}t = \frac{L}{ne} \int i(t)\mathrm{d}t, \tag{3.1}$$

while $\mathcal{E}(t)$ is given by

$$\mathcal{E}(t) = \frac{L}{ne\mu} i(t). \tag{3.2}$$

Hence $\mathcal{E}(x)$ can be determined from a plot of $\mathcal{E}(t)$ versus $x(t)$.

Figure 3.6 shows the spatial field profile at the light-entry side of a p-i-n-type solar cell. Note that the field decay is approximately exponential and given by $\mathcal{E}(x) = \mathcal{E}_0 \exp(-x/L_0)$ with $\mathcal{E}_0 = 60\,\mathrm{kV\,cm^{-1}}$ and $L_0 = 2000$ Å. This

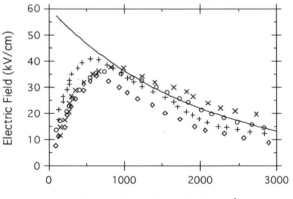

Fig. 3.6. Electric field profiles at the p–i interface in a-Si:H solar cells. The various profiles are due to different layouts of the light entry side of the cell [136]. The solid line shows an exponential decay, given by $\mathcal{E}(x) = \mathcal{E}_0 \exp(-x/L_0)$ with $\mathcal{E}_0 = 60$ kV cm^{-1} and $L_0 = 2000$ Å [156]

behavior contrasts to the typical case of crystalline semiconductors, where the field profile is linear in x, $\mathcal{E}(x) = \mathcal{E}_0(1 - x/L_0)$. The reason for the difference lies in the large density of mid-gap states in the amorphous material, which determine the space-charge distribution in the depletion region. It can be shown that an exponential field profile is consistent with a constant mid-gap DOS. The decay length, L_0, depends on the DOS, and an evaluation shows that for the case of Fig. 3.6, $g(E) \approx 10^{16}$ cm^{-3} [147].

3.2.4 Transition From Shallow to Deep States

Shallow trapping in the band-tail region dominates the transport only in the initial time regime. Eventually, when the band-tail region is thermalized, the mid-gap states also start to participate in the carrier traffic. These will immobilize the captured carriers for much longer times, since they are considerably deeper in energy. Applying (2.9) to estimate the residence time in states near mid-gap gives times of the order of seconds and larger. Thus, only when the time frame of the experiment is extended beyond the 1 s range will the emission from these deep states become apparent. For shorter experiments the carriers are immobile. The cross-over from shallow to deep trapping will occur when the capture probability of the deep states exceeds that of the shallow states. With σ denoting the capture cross-section, N the state density, and the subscripts s and d referring to shallow and deep states, the transition in the trapping behavior occurs at a time τ, given by [157]

$$\tau = \nu^{-1} \left(\frac{\sigma_s N_s}{\sigma_d N_d} \right)^{1/a}. \tag{3.3}$$

Carrier capture into deep states can be studied quantitatively in a modification of the time-of-flight experiment where the time integral of the transient photocurrent is measured. In these so-called charge-collection experiments, the dependence of the charge on the applied electric field is evaluated. When capture by a deep state takes time τ, the probability for a generated carrier to reach the back contact is $p = \exp(-t/\tau)$. Accordingly, the charge collected at the rear contact depends on the applied field [158] as,

$$Q(\mathcal{E}) = Q_0(\mu\tau\mathcal{E}/L)\{1 - \exp[-L/(\mu\tau\mathcal{E})]\}, \tag{3.4}$$

where L is the film thickness and Q_0 is the photogenerated charge. The experiment is useful mainly for the characterization of unipolar transport, which typically occurs in xerographic applications, charge storage devices and transistors. Figure 3.7 shows results of charge collection experiments carried out on a-Si:H photodiodes, where knowledge of the $\mu\tau$ product is used to determine the capture cross-sections and the trap densities [147].

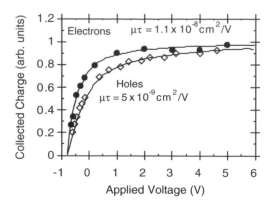

Fig. 3.7. Results of charge collection measurements on a-Si:H solar cells. Trapping $\mu\tau$ products for electrons and holes are obtained [147]. Charge collection begins at negative applied bias due to the built-in potential of the solar cell

3.2.5 Trapping in Deep States

In samples with sufficiently high mid-gap densities, trapped carriers may give rise to strong space-charge fields, which can easily be probed in photoelectric response measurements. Typical experiments involve two laser pulses, one trap-filling pulse and a second probing pulse, whose delay is varied to determine the temporal evolution of the trapped charge. In a-Si:H and in a-SiC:H alloy films trap depths may exceed 1 eV, and excess carriers can be stored for extremely long times [159,160]. Figure 3.8a shows a set of data obtained for a-SiC$_x$:H alloy films with x in the range $0 < x < 0.35$. It is seen that charge retention can easily reach the scale of hours or days in these samples. The long storage times are primarily due to very low recombination rates. These are a consequence of a strong built-in field at the Pt/a-SiC:H interface which provides sufficient carrier separation such that recombination is eliminated. Furthermore, since thermal carrier densities are also low, recombination with

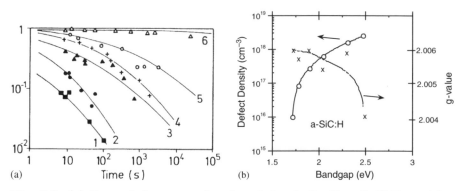

Fig. 3.8. (a) Trapped charge as a function of time in 6 a-Si$_{1-x}$C$_x$:H films with x increasing from 0 (■) to 0.35 (△) [153]. (b) Defect density in a-SiC:H determined by ESR. For increasing C content of the film the g value shifts from 2.0055 for the dangling Si bond towards 2.003 for the dangling C bond [159]

thermal carriers is also negligible. From the temperature dependence of the charge retention and a simple model for the thermal-emission mechanism, one is able to determine the trap depth and the trap distribution [159].

The evaluation shows that the trap depth and the trap density increase with C content. Figure 3.8b shows the spin density and spin g factor measured in ESR measurements on the alloyed films. The data suggest that C and Si dangling bonds are the dominant defects in these films [161]. Since trapped charge density and spin density have approximately the same value, it is concluded that the trapping centers are dangling bonds.

3.2.6 Correlation Between Shallow and Deep States

Since the shallow states in the tails of the valence and conduction bands arise from strained bonding configurations in the amorphous network, it is natural to ask about their correlation to the deeper-lying mid-gap states, which arise from broken Si bonds. Several researchers have addressed this question [162–164], and the expected correlation has indeed been established for a large range of samples.

Figure 3.9 shows data for a-Si:H, including undoped, doped, annealed, intentionally contaminated and various other types of films. The dangling bond density is plotted versus the Urbach parameter, E_0, which characterizes the exponential region of the optical absorption edge. In the present context, the data provide a useful basis by which to determine the range of transport parameters in a-Si:H films. On the basis of the multiple-trapping model, with parameters as given in Table 3.1, one obtains for the electron and hole drift mobility ranges

$$10^{-4} \text{ cm}^2(\text{Vs})^{-1} < \mu_{\text{dc}} < 2 \text{ cm}^2(\text{Vs})^{-1}$$

and

$$8 \times 10^{-6} \text{ cm}^2(\text{Vs})^{-1} < \mu_{\text{dh}} < 2 \times 10^{-2} \text{ cm}^2(\text{Vs})^{-1},$$

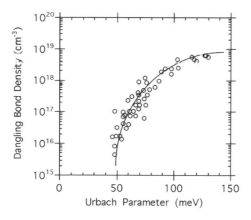

Fig. 3.9. Correlation between dangling-bond density and Urbach parameter. Data are taken from [162]

while the expected ranges for the trapping $\mu\tau$-products are given by

$$4 \times 10^{-10} \text{ cm}^2(\text{Vs})^{-1} < \mu\tau_{\text{e}} < 3 \times 10^{-8} \text{ cm}^2(\text{Vs})^{-1}$$

and

$$3 \times 10^{-11} \text{ cm}^2(\text{Vs})^{-1} < \mu\tau_{\text{h}} < 3 \times 10^{-9} \text{ cm}^2(\text{Vs})^{-1}.$$

Note that the parameters determined for solar cells in the previous sections lie close to the upper limits of these ranges.

3.3 Unipolar Devices

3.3.1 Light-Activated Charge Storage Device

The large retention times discussed in Sect. 3.2 can be exploited in a light-activated charge storage device [165], which operates on a sub-nanosecond time scale. The device consists of a Pt/a-SiC:H/Al sandwich structure with a semi-transparent Pt contact at the light-entry side. In the writing process the sample is illuminated in a scanning fashion using a sequence of laser flashes or with a single flash containing an image. To enhance the charge separation in the writing process, it is sometimes useful to apply an external field. The probing laser flash used for read-out is applied in scanning fashion in short circuit conditions. The corresponding read-out signal is a short (≤ 1 ns) electrical current pulse with an amplitude proportional to the trapped space charge. For multiple read-out it is necessary to decrease the intensity of the probing flash to a level that leaves the read-out charge small in relation to the trapped charge. In practical cases several 100 read-out processes can be obtained with a 1 mV signal amplitude. Other read-out modes are possible when an electric field is applied, or when the stored charge is replenished. Erasing of the stored charge can be achieved by uniform illumination of a few seconds using a dc light source or by a mild heat treatment.

Since the a-SiC:H films are x-ray sensitive, applications include digitized x-ray images for medical and technical purposes, beam position sensing, etc.

3.3.2 Thin-Film-Transistor (TFT)

A-Si:H field effect transistors are widely used for pixel control in active matrix-addressed flat panel displays. The main requirements for these applications are low leakage currents in the off-state, switching speeds compatible with the $\sim$30 Hz addressing cycle and high on-currents. In material terms these demands translate into requirements for a low defect density and high carrier mobilities. Due to the superior electron-transport properties, most devices are laid out for operation in the n-channel accumulation mode. Since

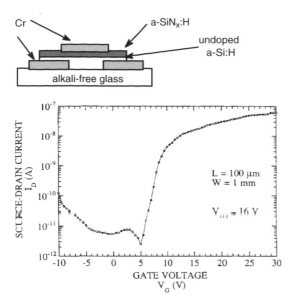

Fig. 3.10. Structure and transfer characteristics of the bottom-nitride a-Si:H transistor prepared in an all-printed patterning process [37]

in the on-state, the channel region is only between 10 and 100 Å thick, a low interface state density near the gate insulator is of great importance [166,167].

Figure 3.10 shows the design and performance of a particularly simple self-aligned TFT, consisting of source-drain contacts deposited on glass, an undoped a-Si:H channel, an a-SiN:H gate dielectric and a metallic gate contact. The fabrication of the device involves an extremely simple xerographic process [37]: The device is prepared on 50-μm-thick glass foils. The etch mask is provided by the xerographic toner material after the a-Si:H-coated foil is processed in a standard photocopier or a laser printer. The demonstration of an all-printed patterning process is thought to be a first step towards low-cost, large-area circuit fabrication [168].

The transfer characteristics of Fig. 3.10 can be evaluated to determine various transport properties and the DOS in the a-Si:H channel region. Because the channel region is quite thin, the electron transport is sensitive to the quality of the channel/gate interface. Experimental parameters determined from transistor characteristics are therefore not always representative of bulk properties. For the device depicted in Fig. 3.10, the electron drift mobility is, $\mu_d = 0.001$ cm^2(Vs)$^{-1}$. This value is at the lower end of the mobility range discussed in Sect. 3.2 and much smaller than typical values obtained for solar cells, for which mobilities up to 2.5 cm^2(Vs)$^{-1}$ have been measured. The low mobility is not due to inferior a-Si:H bulk properties, but instead results from a broadened conduction band-tail at the nitride interface. Careful optimization of the interface preparation conditions yields drift mobilities up to 1 cm^2(Vs)$^{-1}$ in TFTs.

3.4 Recombination-Limited Transport

The kinetic behavior of excess carriers strongly depends on the occupation of localized states in the gap [157,169]. When only one carrier type is present at significant densities, trapping determines the carrier kinetics. The time-of-flight experiment discussed so far is carried out on reverse biased diodes, i.e., on depleted devices, and the observed kinetics therefore reflect the trapping behavior.

Recombination requires the availability of both carrier types, since it is a process in which the two carrier types have to be captured at the same site. Typical experimental situations are given in photoconductance measurements with both carrier types homogeneously distributed across the film. Differences between trapping and recombination behavior become apparent in the majority-carrier $\mu\tau$ product, when it is determined either in time-of-flight or in steady-state photoconductance experiments. In undoped a-Si:H the two values typically differ by a factor 100, steady-state photoconductance giving the larger value [170].

Although the kinetic behavior in the case of a continuous DOS distribution as in a-Si:H may be quite complicated [171,172], important features can be understood using simple approximations. If one assumes only one type of recombination center in the gap, the recombination times for free electrons and holes are given by

$$\tau_{\rm re} = \tau_{\rm oe} + (n/p)\tau_{\rm oh} \quad \text{and} \quad \tau_{\rm rh} = \tau_{\rm oh} + (p/n)\tau_{\rm oe}, \tag{3.5}$$

where $\tau_{\rm oe}$ and $\tau_{\rm oh}$ are the free-carrier trapping times for electrons and holes, and n and p are the respective free-carrier densities. Note that (3.5) is consistent with the neutrality requirement, which imposes the recombination rate of both carrier types to be equal,

$$R = n/\tau_{\rm re} = p/\tau_{\rm rh}. \tag{3.6}$$

According to (3.5) the recombination time can be many times larger than the carrier trapping time, because the carrier with the longer recombination time can undergo many trapping and re-excitation cycles before the recombination loop is closed. The relevant parameter in this situation is the ratio of the two free-carrier densities.

For doped material, (3.5) predicts that recombination and trapping times for free minority carriers are approximately equal, while for free majority carriers the recombination times may be much larger than the trapping times. Accordingly, in doped material comparably large $\mu\tau$ products for majority carriers are expected.

As was pointed out earlier, undoped a-Si:H is slightly n-type and most of the defect states lie in the lower half of the bandgap. Even in photoconductance experiments with large excess carrier densities, the free-hole density is therefore smaller than the free-electron density. It then follows from (3.5) that the free-electron recombination times are larger than the trapping times, while the free-hole trapping and recombination times are approximately equal. The same holds, of course, for the electron and hole $\mu\tau$ products, which explains the observed differences between trapping and recombination experiments.

Note that (3.5) and (3.6) are based on the free-carrier parameters to be consistent with the definitions in Sect. 2.2. If the shallow carrier densities and the corresponding trapping times are inserted instead, the equations directly yield the recombination times for the shallow carriers, which are the parameters accessible in most experiments.

3.4.1 Undoped a-Si:H

For the determination of recombination $\mu\tau$ products, several different experiments have been employed, all of which involve ambi-polar transport conditions. The results have been derived in the approximation that holes are the minority carriers, which, due to the asymmetric defect distribution in intrinsic a-Si:H, holds to rather high excitation levels.

Steady-state photoconductance measurements have been used in conjunction with the moving-grating technique to study planar sample structures. The photoconductance is given by $\Delta\sigma = Ge(\mu_{oe}\tau_{Re} + \mu_{oh}\tau_{Rh})$. Since in a-Si:H the transport is dominated by electrons, the electron $\mu\tau$ product is approximately given by $\mu\tau_e \approx \Delta\sigma/Ge$. The moving-grating method [173,174] additionally allows a determination of the drift mobilities for both carrier types and the ambi-polar recombination time (for the shallow carriers), such that the hole $\mu\tau$ products can be determined for the same samples. For sandwich structures the surface-photovoltage method was employed to determine the ambi-polar diffusion length, L_D, which, within the approximation, equals the hole diffusion length. Use of the Einstein relation then gives the recombination $\mu\tau$ product as $\mu\tau_h = eL_D^2/kT$. The junction-recovery method has been used for a determination of hole recombination $\mu\tau$ products in pin devices [113]. Figure 3.11 shows a summary of results for a wide selection of undoped a-Si:H samples using these methods. The results support the expectation that in undoped material the electron $\mu\tau$ product is considerably larger than the hole $\mu\tau$ product, and, furthermore, that for electrons the recombination $\mu\tau$ products are larger than those for trapping. Although there is considerable scatter in the data, it is also apparent that the $\mu\tau$ products scale inversely with the mid-gap density of states, indicating that recombination involves mid-gap centers.

3.4.2 Doped and Alloyed Films

In undoped material the larger electron $\mu\tau$ products are the result of electrons being majority carriers. As expected from (3.5), the dominance of electrons in the transport behavior also holds in more strongly doped n-type material.

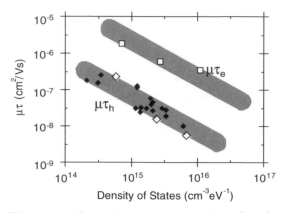

Fig. 3.11. Recombination $\mu\tau$ products for electrons (*squares*) and holes (*diamonds*) in undoped a-Si:H determined from junction-recovery ($\Diamond$) [113], surface-photovoltage ($\blacklozenge$) [118], and moving-grating experiments ($\square$) [173]. The shaded areas indicate the correlation between $\mu\tau$ product and mid-gap DOS

However, with rising doping level, the hole recombination time further decreases, as an increasing number of negatively charged mid-gap states become available for recombination. Thus, the asymmetry in the transport behavior is further emphasized with n-type doping, and the electron $\mu\tau$ products become significantly larger than the hole $\mu\tau$ products. In p-doped a-Si:H the roles of the two carrier types are reversed. Since the electrons constitute the minority carriers, their lifetime is decreased and the $\mu\tau$ product eventually falls below that of the holes. Fig. 3.12 summarizes results from junction-recovery experiments and compares the results to those of others.

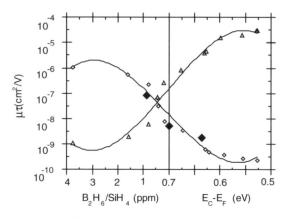

Fig. 3.12. Electron and hole $\mu\tau$ products for p- and n-doped a-Si:H films. Data are taken from Yang et al. ($\Diamond$) [175], Kocka et al. ($\triangle$) [169], and from junction-recovery experiments ($\blacklozenge$) [113]

Alloyed Samples. Much effort has been directed toward the optimization of the electronic properties in a-Si:H alloys. The most promising results so far have been obtained with a-SiGe:H, which has a smaller band gap than a-Si:H, and with a-SiC:H, which has a larger gap. It must be admitted, however, that the transport properties suffer significantly with alloying. The general trend in the alloys is that both mobilities and lifetimes decrease with alloy content [33,176]. This has mostly been established by use of transient techniques under trapping conditions, but the same trends also apply in recombination conditions. Figure 3.13 shows results from moving-grating measurements on a set of a-SiGe:H and a-SiC:H samples [173,174]. More detailed discussions of carrier transport in these alloys can be found in [33].

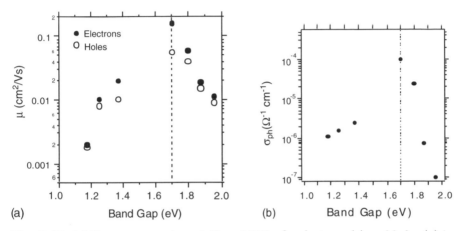

Fig. 3.13. (a) Room-temperature drift mobilities for electrons (•) and holes (○) in a-SiC:H and a-SiGe:II alloys obtained by the moving-photocarrier-grating technique [173]. (b) Photoconductance versus band-gap in a-SiC:H and a-SiGe:H alloys

3.4.3 Structural Relaxation Effects

As pointed out in Sect. 1.2.1, the atomic structure in a-Si:H responds to changes in the electronic system by a slow, observable relaxation, which also involves the defect structure. Light-induced structural changes became known as the Staebler–Wronski effect [177], and are due to an increase in the dangling-bond density [178]. Since the dangling bonds are efficient recombination and trapping centers, the relaxation effects strongly affect the dark conductance and photoconductance. The light-induced effects were soon complemented by findings of relaxation effects induced by carrier injection, thermal treatment, electron irradiation and particle bombardment. As of today, a large amount of experimental data have been collected, but there are still a number of models under discussion [124,131,132].

Figure 3.14 shows that dark conductance is a convenient tool to monitor the relaxation and annealing kinetics. A set of n-type a-Si:H samples was subjected to various thermal and radiation exposures and subsequently annealed at $T = 380$ K. The corresponding changes in the dark conductance exhibit a stretched-exponential time law, $\Delta\sigma \sim \exp[-(t/\tau)^\beta]$, indicative of a broad distribution of time constants for annealing [179,180].

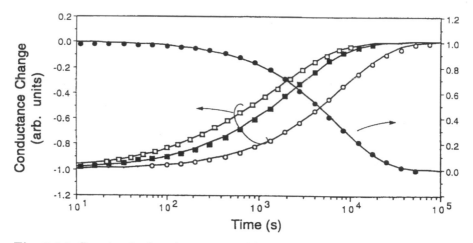

Fig. 3.14. Structural relaxation monitored by conductance measurements in illuminated ($\square$), electron- ($\blacksquare$) and proton- ($\circ$) irradiated, and thermally treated ($\bullet$) n-type a-Si:H samples [32]. Fits based on a stretched-exponential time law are shwon (*solid lines*)

When device-grade material is subject to long-term illumination, as in solar cells, the carrier ranges degrade, while the carrier mobilities remain largely unaffected. This is consistent with the finding of an increased dangling-bond density and comparatively small changes in the band-tail DOS. Figure 3.15 shows MPG results obtained on undoped a-Si:H using a generation rate of 10^{21} cm^{-3} s^{-1} for the degradation. The recombination time decreases by a factor of $\sim$10 in the course of illumination, while the electron and hole drift mobilities remain the same.

3.5 Bipolar Transport: a-Si:H Solar Cells

The basic a-Si:H solar cell is a p-i-n structure of $\sim$0.5–0.6 μm thickness. The cell thickness is a sensitive parameter that needs to be optimized to allow both, high absorption and satisfactory charge collection. To keep the

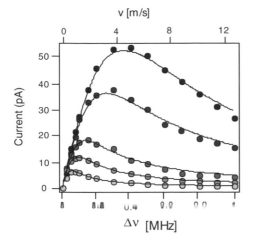

Fig. 3.15. MPG results illustrating the degradation of transport parameters in undoped a-Si:H under long-term illumination up to 76 h at a generation rate of 10^{21} cm^{-3}s^{-1}. Calculated fits to the data (*solid lines*) [173,174] give $\mu_{de} = 0.16$ cm^2(Vs)$^{-1}$ and indicate that τ_{no} decreases from 3 μs to 0.4 μs in the course of the degradation

thickness to a minimum, the devices have a reflective back contact and are deposited on roughened substrate to maximize internal reflection.

Solar cell operation is based on the separation and extraction of photo-excited carriers; the discussion below will illustrate how the understanding of both trapping and recombination kinetics is useful for the optimization of the device performance. Due to the small minority carrier recombination time in doped a-Si:H, the thickness of the doped contact regions is kept close to a minimum, sufficient only to avoid carrier depletion. Typical thickness values for the doped layers is 100 Å. Absorption, carrier generation and transport occur predominantly in the undoped layer. Since undoped a-Si:H is slightly n-type, the p–i junction usually has a higher built-in potential than the i–n junction. It is therefore often advantageous to place the p–i junction at the light-entry side of the device, resulting in the following device structure: glass/TCO/pin (a-Si:H)/metal. To illustrate the essentials of the carrier traffic in such a device, Fig. 3.16 shows a plot of the free-carrier densities for strongly absorbed light entering a typical cell through the p-side. The data are taken from a calculation by Hack et al. [181,182]. It is noted that the density of free holes is strongly peaked near the light-entry side, and drops steeply towards the rear n-contact, while the free-electron density is distributed more evenly throughout the device. Since the recombination rate is proportional to the product of the free-carrier densities, the steep hole profile implies that recombination also peaks near the p-contact. The calculations indeed show that the recombination rate drops by more than two orders of magnitude within 0.4 μm from the front surface. The distributions in Fig. 3.5 indicate that the two carrier types do not play a symmetrical role even in very thin devices. Thus, the problem to be solved is to optimize the device for the limiting carrier type. The simulation further indicates that true recombination-limited conditions are only encountered in the front portion of

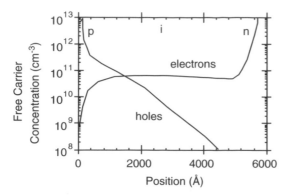

Fig. 3.16. Trapped- and free-charge densities in the a-Si:H solar cell, illuminated through the p-side (data are taken from [181,182])

the cell, while trap-limited, majority-carrier transport dominates towards the rear contact. Since electrons have larger mobilities, it is clear that electrons should take the role of the limiting carrier. This allows faster extraction from the recombination region and smaller space-charge build-up in the transport region. The superior transport properties of electrons also suggest that they are collected at the more distant contact, which is the non-illuminated rear contact. Light entry from the p-side thus appears to combine a number of advantages for a-Si:H solar cells.

Although one would expect inferior device efficiencies for glass/nip/metal structures, efficient devices have also been fabricated with this geometry. This is understood by considering that it is mostly the $\mu\tau$ product that determines the conversion efficiency. Due to the strong sensitivity of the $\mu\tau$ product to the Fermi-level position (Fig. 3.12), very small doping concentrations, typically below the 1 ppm range, can improve the hole $\mu\tau$ products and make them equal to or better than the electron $\mu\tau$ product. Hence, when the center region is lightly p-doped, devices illuminated from the n-side can perform at a similar efficiency as those with p-side illumination. Boron-doping of the absorber layer has led to good results in several cases. However, the incorporation of low-level doping into the fabrication process involves a serious technical complication and has not been pursued on a large scale.

Today's optimized a-Si:H solar cells comprise a number of technical improvements on the simple design discussed here. The most important of these involve bufferlayers between the front and rear contact regions, crystallized doped layers, a-SiC:H window layers, hazed TCO front contacts, and reflecting and optically optimized back contacts. With these improvements the peak efficiencies of single-junction cells have reached 13%; stabilized efficiencies are around 11%.

3.6 Summary

Localized state effects dominate the electronic transport behavior in a-Si:H. In most applications the continuous distribution of gap states and the spatial variation of their occupation make it necessary to employ numerical techniques to describe the transport. Only a semi-quantitative description can be attained by employing several simplifying approximations. Of these approximations, the distinction between a set of shallow, thermalized states and a set of deep states with non-thermalized occupation is one of the most useful. Qualitatively it can be stated that the shallow states determine the carrier drift mobility and affect the temporal response of devices up to the μs region. Deep states act as recombination centers or as deep traps with very long residence times. For a given doping level, the carrier ranges are approximately inversely proportional to the density of deep states, but the carrier ranges vary strongly with doping. Even in device-grade material with low gap state densities, carriers in deep states determine the steady-state screening behavior in devices and limit the extension of space-charge regions.

4. Transport in Porous TiO₂ and TiO₂-Based Devices

4.1 Basic Characterization of Porous TiO₂

A sensible structural description of porous films requires a large set of parameters, such as grain size, porosity, pore geometry, connectivity, crystallite shape, grain-surface, structure etc. All of these parameters may strongly influence the physical properties of the films and thus may influence whether the fabrication of multi-layer structures and devices is possible. Due to the difficulty in studying such a large set of parameters, a more heuristic approach is taken in the following, emphasizing those results and conclusions which are useful for device development.

4.1.1 Structural Characterization

Figure 4.1a shows an x-ray diffractogram obtained on thin films of anatas-phase porous TiO_2. The strongly broadened peaks are attributed to the small grain size of the nano-crystalline clusters. Applying Scherrer's analysis for an estimation of the average cluster diameter gives $d \approx 60$ Å.

Figure 4.1b addresses the question of thermal stability of the porous films. Differential thermal analysis indicates a phase transition to occur between 650°C and 750°C. In this transition the anatas phase is transformed into rutile. As in the crystal, the rutile phase is the stable phase and is maintained after subsequent cooling. In bulk TiO_2 crystals, the anatas–rutile transition is observed at approximately the same temperature. X-ray analysis indicates that the phase transformation is accompanied by substantial grain growth (Fig. 4.1c).

4.1.2 Doping

In TiO_2 crystals, n-type doping is achieved by changing the Ti/O ratio or by doping with electron donors, such as V, Cr, and Nb. By both methods large increases in the electron concentration can be obtained. In nano-crystalline films impurity doping appears to be far less effective [183,184]. Apparently only a very small fraction of the impurities form substitutional sites; it appears instead that the dopants are themselves fully oxidized by

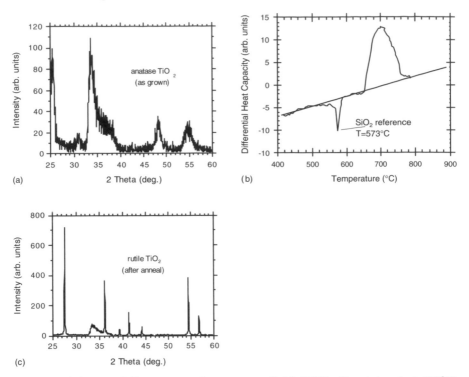

Fig. 4.1. (**a**) x-ray diffractogram of as-grown colloidal TiO$_2$ film sintered at 450°C, showing broadened peaks of anatas phase [39]. (**b**) Differential thermal analysis results on colloidal TiO$_2$ films, indicating a phase transition from anatas to rutile at $T \approx 650°$C [39]. (**c**) Same film as in (**a**) but annealed at 1100°C, indicating larger grains in the rutile phase [39]

incorporating additional O into the lattice. Only at very high impurity concentrations conductance increases are observed, but it appears that these are due to defect-assisted hopping transport. Li intercalation appears to be another method of impurity doping that may lead to significant conductance changes in the nano-porous films. Li is placed at an interstitial site in the TiO$_2$ lattice and induces the creation of Ti^{3+} states [185,186]. The intercalation can be achieved by chemical or electrochemical methods [187,188]. Insertion ratios up to $x = 0.8$ in Li$_x$TiO$_2$ have been achieved, which makes the Li$_x$TiO$_2$ system also a candidate for battery [189] and electrochromic applications [190]. However, quantitative results on the induced conductance changes have so far not been reported.

The most effective way of doping the nano-crystalline films appears to be via the formation of O-vacancies by means of non-stoichiometric preparation, or by reduction in post-deposition processes [39,191]. Due to the extremely large surface area, only small transport distances in the bulk are necessary to bring oxygen to the surface. Hence, when O is desorbed from the surface, concentration gradients within the film will be equilibrated on a fast time scale, and bulk-like doping is readily achieved.

Insight into possible defect reaction mechanisms for oxygen-deficient TiO$_2$ can be obtained from detailed investigations on single crystals. Two different mechanisms for doping based on oxygen deficiencies have been suggested. These involve the formation of anion vacancies and the formation of Ti interstitials [192]. A simple chemical model considers the lattice to be made up of Ti^{4+} and O^{2-} ions and describes the oxygen deficiency in terms of neutral as well as singly and doubly charged anion vacancies, V_A. This equilibrium can be expressed as

$$2\mathrm{Ti}^{4+} + \mathrm{O}^{--} \leftrightarrow V_A + \tfrac{1}{2}\mathrm{O}_2(\mathrm{gas}) + 2\mathrm{Ti}^{3+}, \tag{4.1}$$

$$\mathrm{Ti}^{?\,!} + V_A^{\cdot} \leftrightarrow \mathrm{Ti}^{4\,!} + V_A^{\cdot}, \tag{4.2}$$

$$\mathrm{Ti}^{3+} + V_A \leftrightarrow \mathrm{Ti}^{4+} + V_A^{--}, \tag{4.3}$$

where (4.3a) describes the production of anion vacancies, the formation of Ti^{3+} on lattice sites and the loss of oxygen from the crystal. Equations (4.3b) and (4.3c) describe the formation of charged vacancies. Conductance changes result when the vacancy and/or the Ti^{3+} energy levels are sufficiently close to the conduction band such that these levels are thermally ionized.

The alternative mechanism, the formation of Ti interstitials, also produces free excess electrons, as described by

$$\mathrm{Ti}^{4+} + \mathrm{O}^{--} \leftrightarrow \mathrm{Ti}_{\mathrm{int}}^{4+} + 4e^- + \mathrm{O}_2(\mathrm{gas}). \tag{4.3}$$

Intrinsic and hole-producing defect reactions are neglected here, since the experimental data always show a strong doping effect. For rutile the experimental results indicate that vacancy formation is relevant at high O$_2$ pressures, while the interstitial model applies at O$_2$ pressures below $\sim$10 mTorr [192]. It is not known to what extent these two models can be applied to nano-porous anatas films, but they can certainly serve as a guideline to the experimental results.

Desorption of oxygen from the nano-porous films can be induced by thermal, chemical and photo-assisted processes [193,194]. Figure 4.2 shows the conductance of anatase TiO$_2$ films when they are exposed to H$_2$ plasma, which readily removes the surface-bonded O. Before exposure the films show a thermally activated dark conductivity of $\sim$10^{-8} (Ω cm)$^{-1}$. The activation energy typically lies between 1 and 1.6 eV. Within the first 40 s of plasma exposure the conductance changes by approximately 6 orders of magnitude to a saturation level of 10^{-2} (Ω cm)$^{-1}$ and the conductivity activation energy decreases to $\sim$0.35 eV. However, upon admission of O$_2$ into the plasma chamber, one finds an immediate return to the initial intrinsic conductivity value, indicating that the created O-vacancies are not stable in a normal atmosphere. To obtain stable doped films, one has to prepare buried layers of TiO$_2$, or adjust the oxygen surface concentration by adsorbing effective reducing agents. The latter possibility is more easily achieved in liquids [195,196],

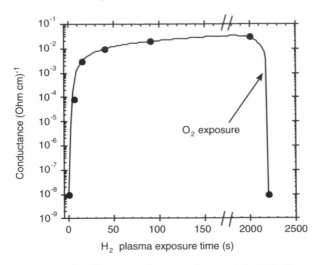

Fig. 4.2. Conductance changes of colloidal TiO$_2$ film exposed to H$_2$ plasma, and upon subsequent O$_2$ admission [39]

which may be an important reason why porous TiO$_2$ films perform well in electrochemical devices [188].

4.1.3 Quantum Confinement

TiO$_2$ films consisting of clusters in the quantum confinement regime can be prepared by carrying out the chemical synthesis at lower temperatures [197,198]. The confinement effects become readily apparent in a blue-shift of the fundamental absorption edge. This is illustrated in Fig. 4.3, which shows the absorption edge for a colloidal ZnO suspension, two porous thin films and a crystalline sample. The main reason for showing spectra for ZnO rather than TiO$_2$ is the steeper absorption edge in ZnO crystals, which greatly facilitates the comparison of the spectra.

The absorption edge of the colloid, consisting of 20–50-Å diameter ZnO clusters in aqueous suspension, is blue-shifted by ~0.2 eV in comparison to the crystalline spectrum. This blue-shift is the signature of quantum confinement in the colloidal crystallites. The colloid can be used directly for the deposition of thin films, but a sintering step is necessary to induce a reasonably stable structure. In low-temperature sintering, the average colloid size is not affected, but the connectivity among the crystallites is increased. This necessarily leads to a reduction of the quantum confinement and a reduced blue-shift of the absorption edge. After sintering at 300°C, the blue shift is reduced to ~0.1 eV. For higher sintering temperatures, the blue-shift decreases further and disappears as temperatures reach 500°C.

Qualitatively similar behavior is also observed in TiO$_2$ [61]. Although the confinement in the porous films is remarkable, we will, in the following, be

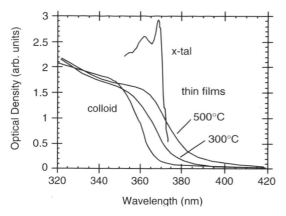

Fig. 4.3 Absorption edge of colloidal ZnO films sintered at 300°C and 500°C, of crystalline ZnO (x-tal), and of a colloidal suspension of ZnO particles (colloid) [39]

concerned with larger grain films, which do not exhibit confinement effects. These films are structurally more robust and can serve as reliable substrates for other semiconductor films and quantum dots. The band-gap of the large-grain TiO_2 films is the same as that of bulk TiO_2, namely ~3.2 eV [199]. The electron affinity is ~4.1 eV [200].

4.2 Photoelectric Properties

The broadening of the optical absorption edge in nano-crystalline films strongly suggests the occurrence of structural disorder, and it is to be expected that disorder effects will affect the photoelectric properties. Figure 4.4 shows the photoconductance decay kinetics in nano-porous TiO_2 films [201,202]. The decay curves have been measured in time-of-flight experiments on Schottky diodes. It is well known that electron mobilities in TiO_2 crystals are much larger than hole mobilities, and it is therefore generally assumed that the photoconductance predominantly involves electron transport [203]. No evidence for a different behavior was found in the porous films, and the results in Fig. 4.4 are therefore attributed to electron transport. The algebraic decay kinetics indicate that trap states and dispersive transport conditions have a strong influence. The dispersion parameter obtained from the slope of these curves is $a \approx 0$, suggesting either a much broader trap distribution than in a-Si:H or a very fast, deep trapping process. The transients in Fig. 4.4 show no indication for complete transit of the carriers through the sample, and the collected charge does not saturate towards high voltages. It is therefore not possible to determine the drift mobility from these experiments.

As in a-Si:H, the kinetics of the trapping process differs from the recombination kinetics, as a slight modification of the experiment is able to

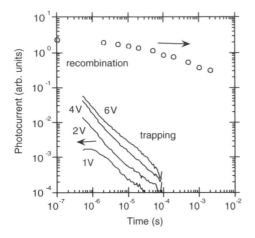

Fig. 4.4. Recombination and trapping kinetics in porous TiO$_2$. Fast photoconductance decay is found under trapping conditions, slower photocharge decay is observed under trap-saturation conditions. In this case the decay is due to recombination [201]

demonstrate very clearly: When the laser pulse is applied without an electric field, the photogenerated carriers are not separated and remain in the thin excitation volume. For sufficiently high intensities, trap saturation will occur and recombination will become the dominant loss mechanism.

For the determination of the decay kinetics, a delayed electric field pulse is used to collect the un-recombined excess carriers. The dependence of the collected charge on the delay then gives information about the recombination kinetics of the shallow carriers. Figure 4.4 shows that recombination of the trapped electrons occurs on a comparably slow time scale.

For a more quantitative investigation of the recombination behavior, the junction-recovery method has been employed. These experiments allowed determination of the recombination time, the $\mu\tau$ product and, calculated from these two parameters, the drift mobility. Figure 4.5 shows the typical current transients from these experiments. It is seen that the recovery process extends over a broad time range, as expected for dispersive transport conditions. The transients show two separate decay regimes, corresponding to drift-limited and emission-limited transport processes. These two regimes exhibit $t^{-0.4}$ and $t^{-1.6}$ time laws, respectively, consistent with a dispersion parameter $\alpha = 0.4$. Apparently, the deep states, which give rise to the extreme dispersion in the time-of-flight experiment, are filled with excess carriers in the junction-recovery experiment, and thus do not contribute to the capture process, allowing the kinetics of the shallow levels to become more apparent.

The value for the dispersion parameter, $\alpha = 0.4$, translates into a band-tail parameter, $E_0 = 62$ meV. Since optical measurements in TiO$_2$ crystals of the anatas phase [203] give very similar values for the Urbach parameter of the optical absorption edge, it is concluded that the DOS near the mobility edge in our films is similar to that in crystals.

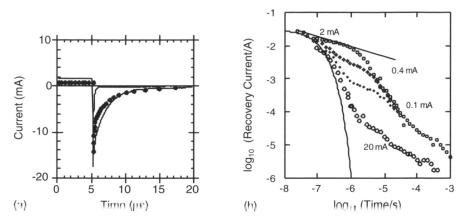

Fig. 4.5. Junction-recovery transients obtained on TiO_2/Pt Schottky barrier devices. The diode is switched from various steady-state forward bias conditions into reverse bias, and the dynamics of the charge recovery is studied to determine the recombination time and the $\mu\tau$ product [114,202]. The film thickness is $\sim$0.5 μm and the device area is 0.1 cm^2. (**a**) linear plot, (**b**) log-log plot

The recombination time for the shallow carriers is obtained from the ratio of stored charge and forward current, $\tau = Q/I_f$ [114], and the $\mu\tau$ product can be calculated from the dependence of the collected charge on reverse voltage. The results of these measurements are shown in Fig. 4.6. Typical $\mu\tau$ values for the nano-crystalline films lie around 10^{-10} $cm^2\,V^{-1}$ independent of the injection level. Since the recombination time decreases with rising injection, the constant $\mu\tau$ data imply that the drift mobility increases at higher current densities. This behavior can be explained in terms of trap filling in the band-tail region. As discussed in Sect. 2.2, the drift mobility is an average taken over all excess carriers, comprising free and trapped carriers, and is given by $\mu = \mu_0 n/n_{\text{tot}}$. When the injection level is raised and the trap states in the gap become filled, the ratio n/n_{tot} is usually not constant [204]. For the case in which the localized states have a band-tail-like exponential distribution, given by $N_0 \exp[-(E - E_c)/E_0]$, it can be shown that, n/n_{tot} is given by,

$$\frac{n}{n_{\text{tot}}} = \left(\frac{\alpha}{2-\alpha}\right)^{1/\alpha} \left(\frac{n_{\text{tot}}}{kTN_0}\right)^{(1-\alpha)/\alpha}, \tag{4.4}$$

where α is the dispersion parameter, E_c the conduction band edge and E_0 the band-tail energy. The dispersion parameter is given by $\alpha = kT/E_0$. Since the exponent $(1-\alpha)/\alpha$ is positive, (4.4) predicts the drift mobility to increase with n_{tot}, i.e., with injection level. At the same time the response time, given by $\tau = \tau_0 n_{\text{tot}}/n$, decreases in a reciprocal manner, such that $\mu\tau$ remains constant. This is exactly what is observed in Fig. 4.6.

In the junction-recovery experiments, many of the parameters for this simple model can be determined experimentally. The dispersion parameter α

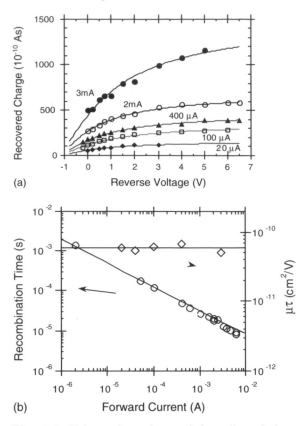

Fig. 4.6. Voltage dependence of the collected charge for various forward current densities, allowing determination of the $\mu\tau$ product; calculated fits are shown (*solid lines*). (**b**) Summary of junction-recovery measurements on TiO$_2$ Schottky diodes [114]. Fits based on the multiple trapping model are shown (*solid lines*)

(and E_0) is obtained from the mobility changes with injection level. As has been shown they can also be determined from the kinetics of the recovery process, just as the $\mu\tau$ product and the response time. The value of n_{tot} is obtained from the recovered charge, Q. The only remaining parameter is N_0, the DOS at the mobility edge. From the theoretical arguments discussed in Sect. 2.1, however, N_0 is estimated to be roughly 10^{22} cm^{-3}. The solid lines in Figs. 4.6a and b represent fits for this choice of N_0 and the parameters given in Table 4.1. The evaluation indicates that the free carrier mobility, μ_0, is ~ 2.4 cm^2(Vs)$^{-1}$, which is similar to what is found in crystals, and further supports the view that the DOS near the mobility edge is not very different from that in larger crystals.

The free-carrier recombination time, τ_0, however, is found to be in the sub-nano-second range, indicating extremely fast capture of free carriers. Since

Table 4.1. Material parameters for TiO$_2$ determined from trap filling model

Band tail parameter, E_{oc}:	62 meV
free carrier mobility, μ_0:	2.4 cm^2(Vs)$^{-1}$
free carrier recombination time, τ_0:	25 ps
DOS at mobility edge, N_c:	10^{22} cm^{-3} eV^{-1}

the crystallite bulk is apparently of good quality, it is concluded that recombination occurs predominantly in surface or interface states.

Potential Fluctuations. The observed mobility changes can also be explained in terms of a different model, based on the screening of potential fluctuations by excess carriers [205]. Statistical arguments, first proposed by Jäckle [206], account for the occurrence of potential fluctuations as a result of localized charges, associated with point defects and impurities [207]. Potential fluctuations are therefore of particular relevance in compensated materials, since these have low thermal carrier densities and a large number of charged impurities. In the following Jäckle's model will be applied to the junction-recovery data and compared to the multiple-trapping picture. For a given impurity concentration, N, and a band-tail carrier concentration, n, the mean amplitude of the potential fluctuations is given by [206],

$$\Delta = 1.28(e/\varepsilon\varepsilon_0)N^{1/3}(N/n)^{1/3}, \tag{4.5}$$

and the screening length for the potential fluctuations is

$$\lambda = 0.28n^{-1/3}(N/n)^{1/3}, \tag{4.6}$$

where $\varepsilon\varepsilon_0$ is the permittivity of the porous TiO$_2$; here a value $\varepsilon \approx 30$ is used. Thus, both the average amplitude and the spatial extension of the potential fluctuations depend on n. Since carriers trapped at band-tail energies have emission times which are short compared to the times which are necessary to establish stationary currents, n has to comprise the carriers trapped in the band tails and free carriers in the band states. The model assumes that carrier transport and recombination involve thermal activation across barriers which are approximately half the size of the average potential fluctuations. The drift mobility and recombination time will then depend on an effective barrier height, $\Delta/2$, as,

$$\mu = \mu_0 \exp(-\Delta/2kT), \tag{4.7}$$

and

$$\tau = \tau_0 \exp(\Delta/2kT). \tag{4.8}$$

The barrier height, $\Delta/2$, thus takes a similar role as the trap depth in the multiple-trapping model. The physical difference between the two models lies in the mechanism of barrier lowering. In the trap-filling model, the average trap depth decreases due to the filling process. This results in a changed ratio

of free to trapped carriers, which in turn leads to an increased drift mobility
for high injection. Since the recombination is transport-limited, the recombination time decreases in reciprocal fashion. In the potential fluctuation
model, the barrier lowering results from screening by the injected carriers,
which also leads to an increase in the drift mobility at higher injection levels
and a lower recombination time.

Since n, μ and τ are experimentally accessible, most model parameters can
be calculated. The solid lines in Fig. 4.7 give an evaluation of the experimental
data in terms of the potential-fluctuation model. The available data give
$\mu_0 = 3 \times 10^{-4}$ cm^2(Vs)$^{-1}$, $\tau_0 = 200$ ns and $N = 1.5 \times 10^{18}$ cm^{-3}.

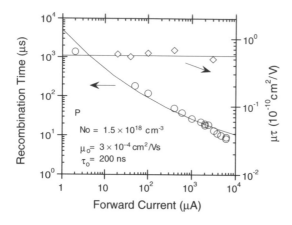

Fig. 4.7. Evaluation of
junction-recovery results in
terms of potential-fluctuation
model. The model parameters
are shown in the figure. Data
are from [114]

The barrier heights lie in the range, 100 meV $< \Delta < 400$ meV, typical screening lengths are between 100 Å and 1000 Å, and typical electric
field strengths due to the potential fluctuations are several 10^4 V cm^{-1}. The
experimental data are reproduced well by the model. Since potential fluctuations are also quite consistent with the exponential tailing of the band edge,
they need to be considered in the transport in nano-porous TiO$_2$. The model
parameters are summarized in Table 4.2.

Table 4.2. Material parameters for TiO$_2$ determined from trap filling model

charge defect density, N:	1.5×10^{18} cm^{-3} eV^{-1}
free carrier mobility, μ_0:	3×10^{-4} cm^2(Vs)$^{-1}$
free carrier recombination time, τ_0:	200 nS
dielectric constant, ε:	30

Probably the most straightforward method to obtain additional evidence for the existence of potential fluctuations is from the current–voltage (I–V) behavior under ohmic contact conditions, using, for example, a planar contact arrangement. One would expect that the application of an external field results in a lowering of the average barrier height in the direction of the current. The field-induced lowering of the barrier would result in an increased drift mobility and thus a superlinear I–V dependence. Figure 4.8 shows I–V plots covering electric fields up to several 10^4 V cm^{-1}.

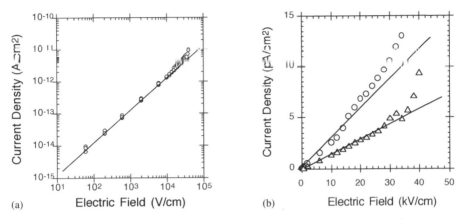

Fig. 4.8. (a) Logarithmic plot of the current–voltage dependence for coplanar strip contacts; strip separation = 500 μm, strip length = 10 mm, film thickness = 5 μm. (b) Linear plot of the current–voltage dependence in the 10^4 V cm^{-1} range for two samples, showing superlinear behavior

Figure 4.8a indicates that below 10^4 V cm^{-1} the dependence is accurately ohmic, while above 10^4 V cm^{-1} there is a clear superlinear rise in the current, which – unfortunately – soon leads to irreversible breakdown in the film. The superlinear regime is shown in a linear plot in Fig. 4.8b. It is noticeable that the superlinearity is accompanied by small irreproducible variations in the current density, which lead to scatter in the data. These temporal variations may be a pre-breakdown phenomenon. While the superlinear behavior is compatible with the existence of potential fluctuations, there are also several other effects, such as space-charge injection, dielectric breakdown and filament-forming effects in the porous network, which can all give rise to superlinear currents. It would be worthwhile to study this behavior in more detail, but it will certainly remain a difficult task to use these data to discriminate between the two transport models. At present it can only be stated that unequivocal experimental evidence for or against potential fluctuations in the porous TiO$_2$ could not be obtained. It is notable, however, that the charged defect density, N, derived from the potential-fluctuation model is surprisingly large. Neither optical experiments nor any other experiments give evidence for the presence of such large, deep defect densities in the nano-porous films. It is also difficult to give a straightforward physical explanation for the low free-carrier mobility suggested by the model and the comparably large free-carrier recombination time. In fact, decay kinetics much shorter than 200 ns have been observed in time-resolved photoconductance measurements, but these fast decays have been attributed to trapping in shallow states rather than recombination in deep centers [201]. For these reasons the trap-filling model may appear preferable, therefore the following picture for carrier propagation in nano-crystalline films is suggested: Similar to what is observed in amorphous films, the carrier transport in the porous films exhibits dispersive behavior [201,208–210]. Electron transport and recombination kinetics can be described in a multiple-trapping model involving an exponential band-tail. While the transport within the grains appears to be fairly efficient, surface and interface states induce very fast capture, which appears to limit the overall transport and recombination behavior. It is to be expected that for device applications surface and interface states need passivation.

Recent work on electrolytically contacted films appears to be quite compatible with these conclusions [208,211–213].

4.3 Solid-State Hetero-Junctions

4.3.1 Schottky Barrier Devices

Figure 4.9 shows the I–V characteristics of a Schottky diode involving a SnO$_2$/TiO$_2$ ohmic back contact and a semi-transparent 80 Å Pt front contact. The I–V dependence shows a well-developed exponential region with an ideality factor, $n \approx 2.5$. Above $\sim$1.2 V ohmic resistance leads to deviations from the exponential dependence, and above $\sim$2 V space-charge injection results in a power-law dependence, $I \sim V^2$. The rectification ratio reaches $\sim$10^4.

It is easily noted in the measurement process that the response to voltage changes is rather slow. To illustrate this point, the inset in Fig. 4.9 shows a typical current transient in response to a fast voltage change. The slow response is a direct consequence of a large density of trap states and low carrier mobilities. The I–V dependence shown in Fig. 4.9 was measured at a sufficiently low scan rate for the steady-state behavior of the device to be obtained.

4.3.2 Semiconductor Hetero-Junctions

a-Si:H/TiO$_2$ Hetero-Junction. Since a-Si:H is deposited in a plasma-CVD process, the a-Si:H/TiO$_2$ hetero-junction is an example of a nearly planar device structure in which the electrical contact is restricted to the uppermost

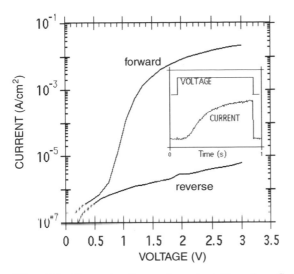

Fig. 4.9. Current-voltage characteristics of an $SnO_2/TiO_2/Pt$ Schottky-barrier-type diode. Exponential forward current behavior, large rectification ratios and small reverse currents are commonly observed in these devices. The inset shows the response to a fast voltage step [201]

surface area of the porous TiO_2 film. The diode studied here consists of the a-Si:H/TiO_2 hetero-junction and a semi-transparent Schottky barrier to the a-Si:H established by a thin Pt layer. The TiO_2 is provided with an ohmic back contact, consisting of degenerately doped SnO_2.

Since most of the internal surface of the TiO_2 is not in electrical contact with the deposited a-Si:H, there are essentially three materials involved in the electrical charge balance within the device: a-Si:H, TiO_2, and the gaseous or liquid ambient adsorbed on the TiO_2 surface. Such three-phase systems are the basis for many sensor applications [193,214]. Figure 4.10 shows that the photo-response of the layer structure is strongly dependent on the ambient in contact with the porous layer. When operated in air, the Fermi level in the porous TiO_2 is close to the mid-gap position. The potential distribution of the layer structure is then as indicated in Fig. 4.10b. Due to the different bandbending at the a-Si:H and Pt interfaces, the current response to blue and red light is of different polarity (curve 1). By immersing the device into a redox electrolyte with a sufficiently low redox potential, the TiO_2 is doped due to electron transfer from the redox pair. In a planar film this type of charge transfer only involves a narrow region at the surface, giving rise to a well-defined space-charge region in the solid and a Helmholtz double layer in the liquid. In the porous structure, however, the volume of the space-charge region is enhanced and in the present case, where the grain size is of the same scale as the space charge width, the charge transfer affects practically all of the TiO_2 layer. In a first approximation, the film can therefore be considered

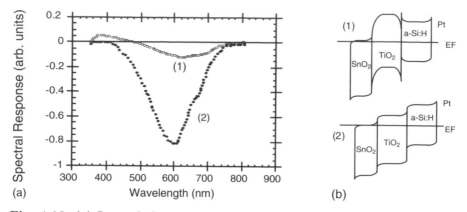

Fig. 4.10. (a) Spectral photocurrent response in an a-Si:H/TiO_2 hetero-junction [199]. Curve *1* is obtained with the device in air, curve *2* with the device immersed in Na_2S/H_2O. (b) Bandbending for curves *1* and *2*

homogeneously doped with electrons. The potential distribution resulting from this doping is depicted in diagram (2) of Fig. 4.10b. It can be seen that the electric field can have the same direction across the whole device, resulting in a photocurrent response of the same polarity at all wavelengths [39,199], as shown in curve 2 of Fig. 4.10a. These results indicate that the a-Si:H/TiO_2 hetero-junction is principally suited for photodetector or solar cell applications when the TiO_2 is sufficiently n-type. The potential distribution across the a-Si:H absorber layer then drives photogenerated electrons to the TiO_2 interface, while holes are transported to the metal contact. The band alignment at the interfaces only allows the transfer of one carrier type and thus promotes the separation of photogenerated carriers. Similar to the Schottky diode, the semiconductor/TiO_2 hetero-junction has an extremely slow current response due the slow trap kinetics in the porous layer. This is illustrated in Fig. 4.11a for a porous device and a compact-layer device prepared by vacuum evaporation of TiO_2. It is noted, that, despite its much slower response, the performance of the porous device matches that of the compact film, supporting the conclusion that efficient transport may occur in the porous layers under saturated-trap conditions. Figure 4.11b shows the *I–V* characteristics of this diode under moderate white-light illumination. The energy-conversion efficiency is found to be in the 1% range for this non-optimized device.

4.3.3 Quantum Dot Hetero-Junctions

Theoretical Background to Confinement Effects. In a potential that confines a small particle in all three dimensions, the DOS becomes a set of bound states at discrete energies. The separation of the levels increases as the spatial extension of the quantum dot is reduced. Several techniques can be employed

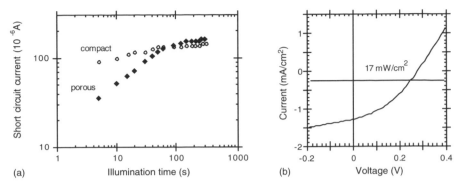

Fig. 4.11. (a) Time evolution of a short-circuit current upon illumination of SnO$_2$/TiO$_2$/a-Si:H/Pt diodes of varying thickness, indicating slow filling of TiO$_2$ traps. (b) Current-voltage characteristics for an SnO$_2$/TiO$_2$/a-Si:H/Pt diode at 17 mW cm^{-2} white-light illumination through the TiO$_2$ side [202]

to give an approximate quantitative description of the problem. The simplest of these is the "electron-in-a-box" model with of the effective-mass approximation. In this approach the Schrödinger equation is solved with a potential that is infinite outside the confinement radius, R. Inside the confined region the potential is given by the averaged Coulomb interaction of the excited electron–hole pairs. The Schrödinger equation then reads [215]

$$\left[\frac{\hbar}{2m_{\rm h}} \nabla_{\rm h}^2 + \frac{\hbar}{2m_{\rm e}} \nabla_{\rm e}^2 + V(r_{\rm e}, r_{\rm h}) \right] \psi = E\psi, \tag{4.9}$$

with $V(r_{\rm e}, r_{\rm h}) = -e^2/\varepsilon|r_{\rm e} - r_{\rm h}|$ inside R, and $V(r_{\rm e}, r_{\rm h}) = \infty$ outside R.

The solution is,

$$\Delta E = \frac{\hbar^2 \pi^2}{2R^2}(1/m_{\rm e} + 1/m_{\rm h}) - \frac{1.8e^2}{\varepsilon R}, \tag{4.10}$$

where ΔE is the band-gap energy. The first term in (4.10) is the kinetic energy of the electron and the hole, which increases as the particle size increases. The second term is the screened Coulomb interaction, which stabilizes the electron–hole pair.

Figure 4.12 shows a plot of the experimental data for the PbS band-gap in the quantum-confinement region. The fit of the "electron-in-a-box" model to the experimental data is not very good. This is mainly due to a breakdown of the effective-mass approximation, which is correct only for wavevectors small in comparison to the reciprocal lattice vector, $k \ll G = 2\pi/a$. The solution, (4.10), has an effective k given by, $k = \pi/R$. Therefore, if k is to satisfy $k \ll 0.01G$, the solution is reliable only for $R > 50a$, that is, for crystallite diameters in excess of ~150 Å. There are also other effects, such as the reduction of the dielectric constant for reduced particle size and surface relaxation, which limit the validity of simple approaches like (4.9). More sophisticated calculations (216–218) have, of course, been carried out,

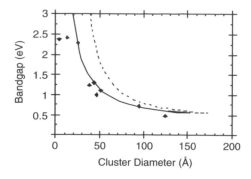

Fig. 4.12. Relationship of effective band gap and cluster sizes in PbS. The data points are experimental results; a band-structure calculation (*solid curve*) and the "electron-in-a-box" model (*dashed curve*) are shown. (data are taken from [217])

and the agreement with the experimental data then becomes much better; this is also illustrated in Fig. 4.12.

The band-gap energy for macroscopic crystals of PbS is 0.41 eV. Quantum-confinement effects set in as the crystal diameter decreases below ~200 Å. The confinement induces a shift in electron energies leading to a sizable increase of the band gap. Since the effective masses in the valence and conduction bands of PbS are approximately equal, the shift of the band-edges is nearly symmetrical. At a cluster size of 25 Å, the opening of the band-gap reaches a value of ~2.4 eV, which corresponds approximately to the energy difference between the ground state and the first excited state in the PbS molecule.

q-PbS/TiO$_2$ Hetero-Junctions. Figure 4.13 illustrates the spatial arrangement of PbS quantum dots in the porous TiO$_2$ layers. The two materials can be distinguished by using Fourier-transform techniques to determine the lattice parameters.

The size-dependent band-gap of the PbS crystals opens new ways for the preparation of hetero-junctions. One important question to be answered in this context is, of course, that of the band alignment at the interface. Figure 4.14 illustrates that bulk PbS imbedded in TiO$_2$ would act as a well for both electrons and holes. Charge transfer from the PbS to the TiO$_2$ would therefore not be possible. For quantum-size PbS, however, the PbS conduction band-edge may approach and pass the TiO$_2$ conduction band-edge, allowing electron transfer into the TiO$_2$. Based on a symmetric opening of the band-gap, a band-gap of ~1.7 eV is estimated to be necessary for electron transfer. According to Fig. 4.12 the corresponding crystallite size is ~30–35 Å. Due to the much lower TiO$_2$ valence band, hole transfer from the PbS is not possible for this case, and the PbS/TiO$_2$ junction would therefore allow charge separation.

The average transport distance for the electron transfer step in this arrangement corresponds approximately to the radius of the PbS crystallites. For these extremely small distances, a very high electron injection efficiency may be expected. Figure 4.15 shows the injection efficiency spectrum of a ~5 μm thick q-PbS/TiO$_2$ film. Due to the broad size dispersion of the crys-

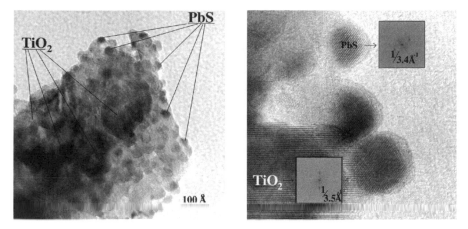

Fig. 4.13. Transmission electron micrograph of PbS clusters on the internal surface of a porous TiO_2 film. In a Fourier-transform analysis of the micrograph (*insets*) the two materials can be identified by their different lattice constants (from M. Giersig, HMI [220])

tallites, no distinct spectral features, such as exciton lines or band-structure singularities, are observed, but it is seen that the TiO_2 is very effectively sensitized by the PbS across the visible spectrum. The quantum efficiency for electron transfer between the two semiconductors reaches a peak value of nearly 80% for photon energies around 2.5 eV. A linear extrapolation indicates that electron injection into the TiO_2 starts at ~1.88 eV, which is slightly higher than the estimate based on Fig. 4.12, and indicates that an excess of ~0.2 eV is necessary to drive the transfer.

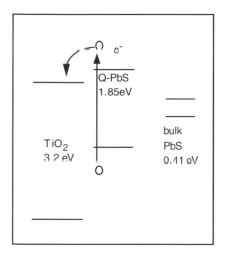

Fig. 4.14. Schematic band diagram of a TiO_2/PbS interface, indicating excited electron transfer from the absorbing PbS to the TiO_2 matrix. On the right, the bandgap of bulk PbS is indicated [44]

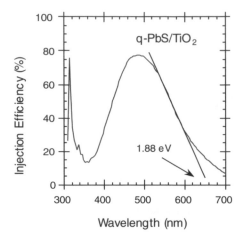

Fig. 4.15. Quantum-injection efficiency for electron transfer from PbS into TiO$_2$ [44]

Figure 4.16 shows a comparison of absorbance and photocurrent spectra. It is noted that the absorption spectra differ quite strongly from the photoresponse spectra. The difference in the blue spectral region is likely due to direct recombination of excited electron–hole pairs, which lowers the electrical response.

In the red region of the spectrum, the long absorption tail is caused by the presence of PbS particles with diameter >30 Å. These absorb in the red spectral region, but cannot transfer the excited electrons into the TiO$_2$ due to their smaller band-gap and lower conduction band. The rise of the photocurrent spectra in the red region thus depends solely on the band alignment.

When the PbS deposition is repeated several times, both the density of quantum dots and the average quantum-dot size increases [61]. Repeated deposition may also lead to a clustering of PbS crystallites. While the photoresponse is expected to increase with the number of quantum dots, clustering and growth will result in a lowering of the transfer probability. Apparently, the latter two effects dominate the behavior in Fig. 4.16, which shows a decreasing photoresponse and increasing absorbance as the PbS deposition is repeated.

The dark conductivity in the PbS/TiO$_2$ films is found to be close to that of the uncoated TiO$_2$, suggesting that the PbS clusters do not form a connected layer. The transport of excess carriers therefore only occurs through the TiO$_2$ and results in a very low dark conductivity of the complete hetero-structure. The low dark conductivity brings about a high photosensitivity, which is important for detector applications.

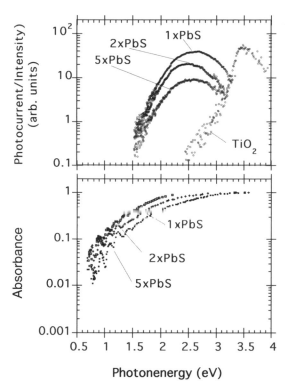

Fig. 4.16. (a) Photocurrent response of TiO_2/PbS films for various numbers of PbS coatings. (b) Absorption spectra of the same TiO_2/PbS films [44]

4.4 Devices

4.4.1 The Eta-Cell Concept

Conventional solar cells are planar devices whose thickness is determined by the requirement that a sizable fraction of the solar photons is absorbed [220]. Materials with high optical absorption coefficients therefore allow thin devices, while materials with low absorption coefficients need comparatively thick absorber layers. In planar cells the photo-excited carriers have to be transported across the total thickness of the absorbing layer, and this requirement generally defines the level to which the electronic properties of the absorber have to be optimized. With a change in the geometry of the cell it is possible to reduce the transport path length in the absorber and thereby relax the requirements for the transport properties, particularly for large $\mu\tau$ products. It is therefore believed that new device geometries and structures may open the way to less expensive devices, allowing simpler processes and a wider range of materials to be employed. The eta-cell presents a novel design for an all solid-state solar cell with an extremely thin absorber [39,199,221].

In this cell the absorber layer is deposited on a highly structured substrate to form a folded layer, as schematically shown in Fig. 4.17a. In this arrangement the transport distance within the absorber layer can be small, while, due to the folding, the optical thickness is sufficient to give a large absorbance. Since these devices present a new advance in the use of low-mobility materials, they are discussed here in some detail. The cells use a structured transparent substrate, which defines the geometry. On top of the substrate layer one deposits a very thin absorber layer in a conformal manner. The total volume of the absorber is given by the requirement that sufficient light is absorbed, while the local thickness of the layer depends on the surface enlargement given by the underlying substrate. The remaining void volume in this structure then needs to filled by a third transparent layer.

This design concept has partly been motivated by a similar structure which uses an organic dye absorber and a redox electrolyte as the void-filling contacting agent [222]. These electrochemical devices have been developed to a point where conversion efficiency is close to ~10%. However, the liquid component is considered to be an impediment for commercialization. Several efforts to develop an all-solid-state device, while maintaining the dye-based absorber have therefore been undertaken [223,224]. The eta-cell concept also intends to replace the organic absorber by an inorganic semiconductor, and to develop an inorganic all-solid-state device.

In the eta-cell the two transparent layers should be of different conductivity type to allow the effective collection of either electrons or holes, while the absorber should be undoped such that electron and holes have similar

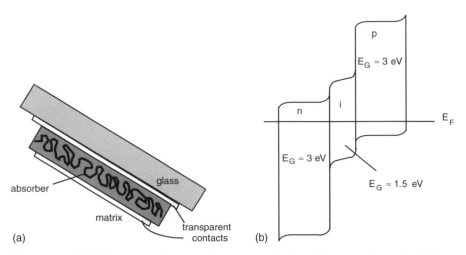

Fig. 4.17. (a) Schematic diagram of a proposed all-solid-state solar cell with an extremely thin absorber. The absorber is prepared as a warped sheet between two wide-gap semiconductor layers. (b) Energy-band diagram of the eta-cell. The absorber with bandgap, $E_G \approx 1.5$ eV is sandwiched between two wide-gap semiconductors with $E_G \approx 3$ eV

$\mu\tau$ products. In most cases this means that the absorber is nearly intrinsic. This reasoning suggests a pin hetero-structure with two active junctions at the p–i and i–n interfaces and a strong built-in electric field across the absorber. The band diagram of this structure is shown in Fig. 4.17b. The valence and conduction band energies of the three layers should be chosen such that they support the separation of the photocarriers generated in the absorber. This means that the conduction band of the n-type contact should lie at slightly lower energies than the absorber conduction band. Furthermore, the p-layer conduction band should should lie at sufficiently high energies, such that electron injection into the p-layer cannot occur. Similar considerations hold for the valence-band alignment in the device. With a band alignment as in Fig. 4.17b, photogenerated excess carriers will be injected as majority carriers into the respective contact layer. While they will still have to cover large distances in these layers, recombination losses will not be severe due to the lack of minority carriers. The requirements for the preparation of this type of structure and the material selection may appear rather challenging, but in the last few years considerable progress towards the realization of these devices has been made [225,226]

4.4.2 Material Selection and First Experimental Results

Figure 4.18 shows a compilation of band-energy data for metal-oxide, phosphide and chalcogenide compounds suited for the eta-cell structure. The data indicate that the metal oxides have comparably low-lying bands [227]. Most oxides are also naturally n-type. TiO_2, ZnO, ITO, WO_3 all have a band-gap of ~3 eV and electron affinities around 4 eV and are therefore natural choices for the wide-gap n-layer. Higher lying bands are found among the II–VI compounds. Among the wide-gap p-type compounds, MnS, ZnTe and MgTe are suited as contacting materials, since they can be prepared from liquid solutions, which makes the preparation compatible with the use of porous oxides as substrate materials. As for the absorber layer, CdTe and CdS have been found to have acceptable conduction and valence band alignment. A similar situation applies for the transition metal dichalcogenides, MoS_2, $MoSe_2$,

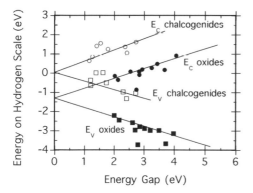

Fig. 4.18. Valence- and conduction-band energies for chalcogenide and phosphide compounds (InP, GaP, Bi_2S_3, CdSe, CdS, ZnSe, GaAs) and oxides (Fe_2O_3, CdO, WO_3, Bi_2O_3, PbO, TiO_2, ZnO, V_2O_5, Nb_2O_5, SnO_2, Ta_2O_5) (data are taken from [227])

and MoTe$_2$. From all available data CdTe appears as a good choice for the absorber between TiO$_2$ and ZnTe contacts. From different considerations we have also proposed CuInS$_2$ as the absorber between n-type TiO$_2$ and p-type CuSCN (226).

Figure 4.19 shows initial results of absorber deposition for the CdTe absorber cell. Conformal coverage of CdTe on the highly structured TiO$_2$ substrate can be achieved using electrodeposition or a novel technique [226] based on a chemical-bath deposition of metal salts with a subsequent gas-reaction step, which produces sulfide or selenide semiconductors such as CdS, CuInS$_2$, CuInSe$_2$, etc. Quite promising device performance is obtained for the TiO$_2$–CdTe junction. Figure 4.19c and 4.19d figure show the photo-electrical characteristics of a device with the structure glass/SnO$_2$/TiO$_2$/CdTe/Au, in which a thin gold film is used as a hole-collecting contact. The analysis of the

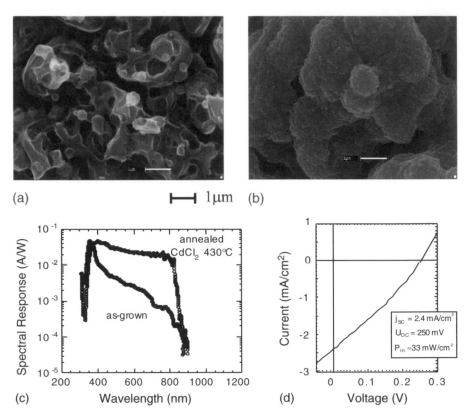

(a)   ⊢━━⊣ 1 µm   (b)

(c)   Wavelength (nm)

(d)   Voltage (V)

Fig. 4.19. Initial experimental results for the eta-solar cell. (**a**) Scanning electron micrograph of the bare nano-crystalline TiO$_2$ substrate film. (**b**) Scanning electron micrograph of an electro-deposited CdTe film covering a nano-crystalline TiO$_2$ substrate. (**c**) Spectral response for CdTe absorber layers in glass/SnO$_2$/TiO$_2$/CdTe/Au-devices. (**d**) Current–voltage characteristics of the same device under 33 mW cm^{-2} illumination

data indicates that the carrier-transfer efficiency across the TiO_2/CdTe interface is approximately 25%; the energy-conversion efficiency for these devices is presently in the 1–2% range. Present effort is directed at the deposition of a suitable p-type contact layer, such as ZnTe, which is expected to improve the performance considerably.

Recent work has also established that the requirement of void-filling deposition can be met using electrodeposition [225]. Figure 4.20a shows, in a sequence of three electron micrographs, how the void volume of a nano-porous TiO_2 film is filled in a continuous-growth mode, starting from the substrate. A combination of weight, coulometric and volumetric measurements show that the TiO_2 pore volume is filled to $100 \pm 3\%$. The current–voltage characteristics of this structure exhibit good rectification, indicating that the TiO_2–CuSCN hetero junction is of reasonable electronic quality. The interface area between the two phases is estimated to be several hundred times larger than the planar projection, and the interface itself is distributed across the whole thickness (~ 8 μm) of the TiO_2 substrate film. The device thus presents a spatially distributed p–n hetero-junction [225]. Completion of the envisioned eta-solar-cell could be achieved by combining both techniques: conformal coverage, as applied for CdTe, and void-filling deposition, as shown for CuSCN. Unfortunately, the deposition processes for these two materials are not compatible with each other. Present efforts therefore attempt to complete the TiO_2–CdTe device by replacing the CuSCN by a p-type ZnTe layer or, alternatively, to insert $CuInS_2$ as the absorber material in the TiO_2–CuSCN hetero-junction.

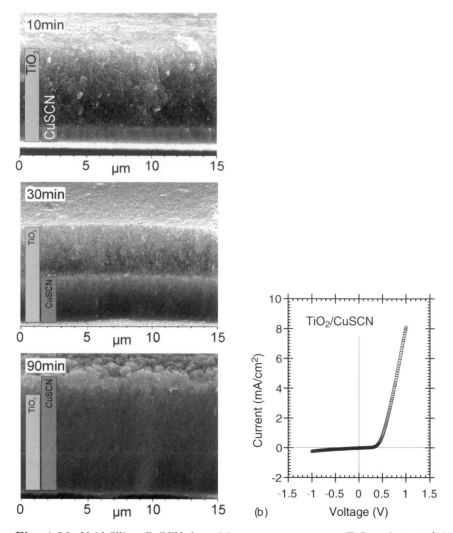

Fig. 4.20. Void-filling CuSCN deposition on a nano-porous TiO₂ substrate [225]. (**a**) The electron micrograph sequence shows the progressing growth front of CuSCN in electrodeposition. The growth front moves continuously from the SnO₂ electrode below the TiO₂, though the nano-porous layer. Void filling is close to 100%. (**b**) Dark current–voltage curve for a TiO₂–CuSCN hetero-junction showing reasonable rectification

5. Transport in C$_{60}$

The inter-molecular forces in solid C$_{60}$ are of the van der Waals type and thus weak when compared to the binding energy of the C$_{60}$ molecule. As a consequence the molecular properties play an important role in the description of the solid, and in many cases the solid state aspects can be understood only with substantial input from molecular physics. This becomes most apparent in the interpretation of the optical spectra of solid C$_{60}$, but is also true for the interpretation of the photo-electric response, as will be seen in the following.

The dominant influence of the molecule on the properties of the solid also simplifies the structural characterization of thin films. As for electron transport, there are mainly two structural parameters of importance – grain size and impurity content. The grain size has already been dealt with in Sect. 1.2.3, purity will become an issue in the context of oxygen content of the films, and will be discussed on several occasions in the following.

5.1 Basic Results

5.1.1 Dark Conductivity

Important basic information on the charge transport in C$_{60}$ can again be obtained from the dark conductivity. Figure 5.1 shows Arrhenius plots of the dark conductance obtained under UHV conditions and after the admission of lab air. In vacuum the conductance of C$_{60}$ is in the 10^{-3} $(\Omega\,\mathrm{cm})^{-1}$ range and is temperature-activated, with an activation energy of $\sim$0.2 eV. Further experiments show that electrons are the majority carriers, i.e., the as-grown films have n-type conductivity, likely due to a charged structural defect or an impurity. When only small amounts of air are admitted to the UHV chamber, the conductance decreases substantially, and the material becomes fairly insulating [228,229]. After long-term exposure to air, the activation energy can reach values of 0.9 eV, indicating that the band-gap is at least 1.8 eV. These strong changes in the conductance are to a large extent reversible. After heating in vacuum to $\sim$300°C, the conductance of air-exposed films shows similar behavior to the as-grown films. There is experimental evidence that the conductance changes are due to oxygen incorporation [230].

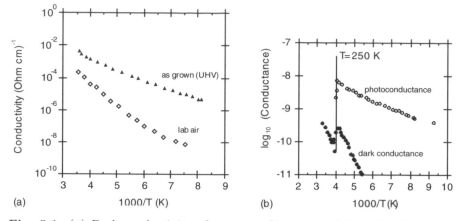

Fig. 5.1. (a) Dark conductivity of as-grown films and oxidized C$_{60}$ films. (b) Discontinuity in the T-dependence of dark and photoconductance at $T = 250$ K due to the structural phase transition (conductance oxygen-exposed sample)

At lower temperatures the conductivity noticeably deviates from the Arrhenius behavior and follows an $\exp(-A/T)^{1/4}$ behavior indicative of variable-range hopping. The deviations from activated behavior occur at higher temperature for the oxygen-free material, suggesting that the localized state density is substantially larger near the donor level than in the mid-gap region. As in a-Si:H the conductance is a sensitive tool for the monitoring of structural changes. Figure 5.1b shows the results for the dark conductance and photo-conductance below room-temperature. The data show a marked discontinuity at ~250 K. This discontinuity accompanies the structural phase transition between the low temperature simple cubic structure and the high temperature face centered cubic structure [79,231]. Apparently, the phase transition also exerts an influence on the electronic system and leads to a somewhat lowered conductance in the fcc phase. The discontinuity is usually not noticed in the more conductive films, since the induced changes are small in comparison to the large dark conductance.

5.1.2 Photocurrent Spectra

The spectral dependence of the photoconductance in C$_{60}$ [232] is shown in Fig. 5.2. The low temperature spectra show a sequence of bands between 1.8 and 4 eV. These bands are usually not found in inorganic semiconductors; they are typical for organic molecular systems and result from a strong coupling between electronic and vibrational excitations within the molecules [233].

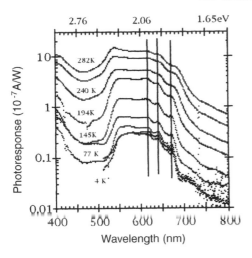

Fig. 5.2. Photocurrent spectra of C_{60}, showing the absorption edge at ~ 670 nm and vibronic structure in the first absorption bands [232]

5.2 Intra-Molecular Excitation Scheme

For an interpretation of the photocurrent spectra the following paragraph will give a brief introduction to the electron levels in C_{60}.

5.2.1 Theoretical Background

The determination of the intra-molecular excitation scheme of C_{60} has been the subject of considerable theoretical effort, and a complete interpretation to the absorption spectral structure has been given [234–238]. An instructive approach to the ground-state configuration is obtained, when one replaces the full icosahedral symmetry of C_{60} by a spherical approximation [239]. This allows the use of spherical harmonics for the classification of the electronic configurations. Since 3 out of 4 carbon valence electrons are consumed in nearest neighbor σ-bonds, there are 60 π-electrons on the C_{60} molecule. In the spherical approximation these arrange themselves into an $s^2 p^6 d^{10} f^{14} g^{18} h^{10}$ configuration. This configuration has zero total angular momentum and is 12 electrons short of a closed shell. When the lower icosahedral symmetry is taken into account, the configuration changes into a ground state with the irreducible representation A_{0g}. In this notation the capital letter refers to the symmetry of the configuration, the first index denotes the excitation state, and g and u refer to gerade and ungerade parity, respectively. The highest occupied state for the A_{0g} configuration is h_u, and the lowest unoccupied state f_u [236]; the lowest excited electronic configuration is therefore $h_u^9 f_u^1$. This excitonic configuration gives rise to singlet ($^1F_{1g}$, $^1F_{2g}$) and triplet ($^3F_{1g}$, $^3F_{2g}$) energy levels. As in many molecules, the triplet levels in C_{60} lie below the singlet levels, but the transition from the singlet ground state into the triplet states is spin-forbidden. Excitation from the ground state into the lowest singlet excited state is symmetry-forbidden, since the nuclear coordinates of both states have gerade parity. In this case, however, the corresponding selection rules can be lifted by Herzberg–Teller coupling [240,241] of vibrational modes of ungerade parity to the electronic interaction. Depending on their symmetry these vibrations are denoted by A_u, F_u, G_u, or H_u. Static structural distortions may also contribute to a lifting of the selection rules.

A schematic level diagram for C$_{60}$ is shown in Fig. 5.3. Electronic levels are drawn as solid lines, while dashed lines represent vibronic levels, resulting from coupling between electronic and vibrational excitations. Some selected transitions are indicated by arrows.

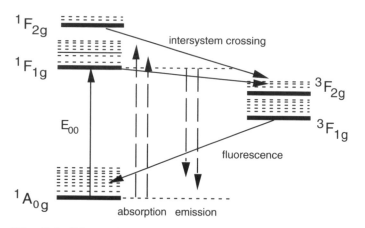

Fig. 5.3. Schematic energy level diagram of vibronic states in C$_{60}$. Vertical arrows indicate the vibronic transitions between the $^{1}A_{0g}$ ground state and the excited singlet state, $^{1}F_{1g}$. Oblique arrows show inter-system crossing transitions and fluorescence

The observed structure in the optical absorption and emission spectra can be satisfactorily explained in terms of the vibronic coupling, illustrated in Fig. 5.3 [242–244]. For most measurement conditions the initial state is thermalized, such that only a few vibrational states are excited. Most of the observed spectral structure is therefore due to vibronic coupling in the final state. In absorption, the initial state is the ground state and the final state is the optically excited state. Hence, the structure in absorption spectra is due to the excited state, and the vibrational energies add to the electronic transition energies. In emission spectra, on the other hand, since the final state is the ground state, the vibrational structure originates from the ground state and the coupling results in a lowering of the transition energies. The vibronic structure in the absorption and emission spectra therefore appears on the high and low energy sides of the vibration-free 00-transition, respectively.

5.2.2 Experimental Results for the 00-Transition

The energy for the vibrationless transition, E_{00}, is usually determined from a comparison of emission and absorption spectra. Since the electronic transition is forbidden, the vibronic structure appears on a low electronic background and is therefore easily resolved. Figure 5.4 shows a comparison of absorption, emission and photoconductance results. Inclusion of photoconductance into this comparison is interesting, since it provides information on the lowest excitation energy for inter-molecular charge transfer. It is seen that photo-conductance follows absorption fairly closely, indicating a constant quantum efficiency at the low electric fields applied in these measurements. Due to the low luminescence efficiency of C_{60}, the vibronic features in the emission spectra could not be observed, and an unequivocal determination of the 00-transition energy based on luminescence and absorption alone was therefore not possible. A tentative assignment, however, can be made from the figure and is $E_{00} = 1.85$ eV. Confirmation of this assignment ultimately comes from the 2-photon luminescence excitation (2-PLE) spectrum [245,246], which is also shown in Fig. 5.4. Since the parity selection rule does not apply in 2-PLE-experiments, the first resolvable peak is expected to arise from the 00-transition. As can be seen, the first peak in the 2-PLE experiments indeed

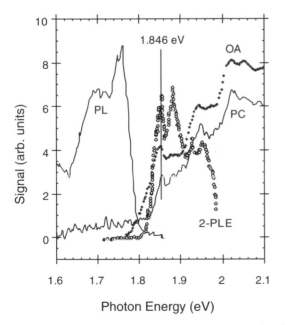

Fig. 5.4. Comparison of optical absorption (OA), photoconductance (PC), photoluminescence (PL) and 2-photon-luminescence-excitation (2-PLE) spectra in C_{60} at low temperatures (4.2–8 K). The data suggest that the 00-transition to lie at 1.85 eV [232,245]

coincides with the 1.85 eV peak in absorption and photoconductance, which establishes the value of E_{00}.

The next step in the spectral evaluation is to correlate the vibronic peak positions with the calculated vibrational energies of the molecule. Since infrared absorption as well as Raman and neutron scattering experiments give complete picture of the vibronic excitation states, it is useful to consult these data for an interpretation of the present photoconductance and absorption spectra. Figure 5.5 shows the results of such a comparison. The peak progression observed in the first absorption and photoconductance bands is explained in terms of two vibrations, ν_1 and ν_2, which are identified as H_u and F_u vibrational modes in the ^1A_g to $^1F_{1g}$ electronic transition [247,248].

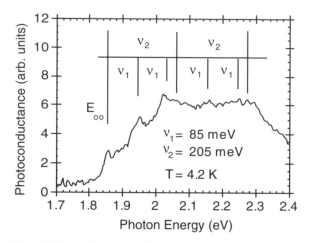

Fig. 5.5. Assignment of peak progression observed in photoconductance spectra, obtained at 4.2 K on thin-film samples [232]

5.3 Intermolecular Charge Transport in C$_{60}$ Films

The question of inter-molecular charge transfer among the low-lying excited states is only partially answered by the observation of photoconductance at photon energies around E_{00}. The comparably weak T-dependence of the spectra in Fig. 5.2 indicates that the transport occurs without the need of large thermal excitation between the final excitation state and a transport state. This may suggest a direct transfer between low-lying excited states as in hopping transport. In this case the excitonic binding energy should, however, be very small, or, alternatively, there should be appreciable state densities near the energies of the lower excited states to mediate the separation of excitons and establish a transport path for excited carriers. If this is not the case, transport can only involve extrinsic mechanisms for excitonic separation, such as ionization at defects or in space-charge regions. From theoretical considerations Louie and Shirley [249] emphasize, that the electron–hole interaction in C_{60} is large and should give rise to strongly localized excitonic levels. This would argue against transport directly among the lower excited states.

We have addressed the question of carrier transfer in several independent experiments. Using pulsed-laser techniques we initially determined the quantum generation efficiency in the energy range from $\sim$1.8 to 4 eV [250]. The experiments establish the probability of generating a separated electron–hole pair after the absorption of a photon. In inorganic semiconductors this probability is often close to one, since excitonic binding energies are small and the bands are sufficiently broad to produce extended state behavior. If the excitonic binding energy is large, on the other hand, local de-excitation can compete with the pair separation. This increases the probability for geminate recombination of the photogenerated carriers and results in quantum generation efficiencies smaller than unity. Other experiments address the strength of solid-state effects in the C_{60} films and aim at the identification of a conduction band in the solid. The data from these experiments allow a precise determination of the band-gap in C_{60} and indicate that the lower excited states have a strong molecular character. The whole set of results appears to indicate that direct charge transfer among the lower excited states does not occur in C_{60}. Finally, from moving-grating experiments, the values of the carrier drift mobilities and the recombination time are determined [88].

5.3.1 Quantum Generation Efficiency and Carrier Range

The idea behind the quantum-efficiency experiments is to generate electron–hole pairs in a thin layer near the surface of the C_{60} film and subsequently collect these by the application of an electric field. The collected photocharge is determined by two parameters, the quantum efficiency and the average drift length of the carriers in the film. Since both parameters depend on the electric field, two complementary experiments need to be carried out for their determination. The first of these is shown in Fig. 5.6a: The sample is subject to a voltage pulse of variable amplitude and is subsequently illuminated by a laser pulse at fixed delay (typically $\sim$5 µs). In this case the quantum efficiency, η, and the drift length of the generated carriers, $\mu\tau E$, change with the applied field, and the photocharge is given by [113,250],

$$Q = eN\eta(E)\mu\tau\{1 - \exp[-L/(\mu\tau E)]\}/L, \qquad (5.1)$$

where e is the electron charge, N the number of absorbed photons, and L the film thickness.

In the second experiment, illustrated in fig. 5.6b, the laser pulse is applied at fixed electric field, and, with a delay of a few ns, a voltage pulse of

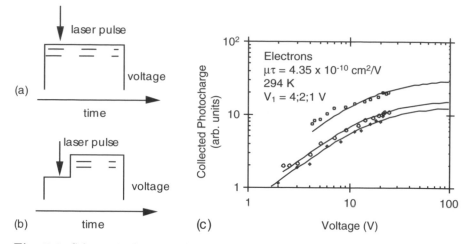

Fig. 5.6. Schematic diagram of charge-collection experiment [250]. (a) Same field for generation and collection, (b) constant electric field for generation and variable collection field, (c) experimental results for C$_{60}$

variable height is applied to collect the carriers which have evaded recombination or trapping. Since the generation step always occurs at the same field, the quantum generation efficiency can be assumed constant. From this experiment, one can therefore determine the mobility-lifetime product using (5.1) and η = const. With knowledge of $\mu\tau$, (5.1) also gives η as a function of field strength. Figure 5.6c shows typical results for the determination of $\mu\tau$. The data exhibit the expected linear field dependence at low voltage, and saturation around 20 V. The applicable voltage range is limited by dielectric breakdown, typically occurring around 4×10^5 V/cm, and the presence of internal electric fields which affect the charge collection below 2 V applied bias. The experimental results can be fitted with (5.1), giving $\mu\tau = 4.3 \times 10^{-10}$ cm^2 V^{-1} for the electron $\mu\tau$ product (under trapping conditions).

The experiments were carried out over a range of temperatures, allowing the temperature dependence of the mobility-lifetime product and the quantum efficiency to be obtained as shown in Fig. 5.7. The experimental results are reasonably well explained in terms of the Onsager model [251,252], which formulates the generation process as a competition between geminate recombination due to the Coulomb attraction of the electron–hole pair and a pair-separation process driven by diffusion and drift. Essentially, when the initial pair separation, r_0, exceeds some critical radius, given by material constants, the pair separates and can contribute to photoconduction. The thermalization radius, r_0, and the transition probability, Φ_0, are the only adjustable parameters of the model. An analytical expression for the quantum efficiency has been derived by Pai and Enck [252],

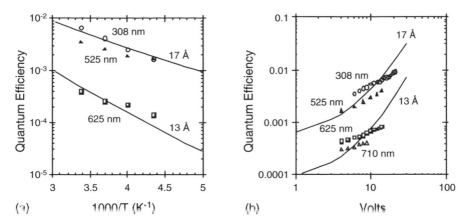

Fig. 5.7. (a) Quantum generation efficiency versus temperature for different excitation wavelengths in C$_{60}$. (b) Quantum generation efficiency versus applied electric field [250]

$$\eta = \Phi_0 \frac{kT}{eEr_0} \exp(-A) \exp\left(\frac{-eEr_0}{kT}\right)$$

$$\times \sum_{m=0}^{\infty} \frac{A^m}{m!} \sum_{n=0}^{\infty} \sum_{m+n+l}^{\infty} \left(\frac{eEr_0}{kT}\right) \frac{1}{l!}, \qquad (5.2)$$

with

$$A = e^2/(4\pi\varepsilon\varepsilon_0 r_0 kT),$$

where e is the electron charge, $\varepsilon\varepsilon_0$ the electric permittivity of the material, and E the electric field. As shown in Fig. 5.7, r_0 typically is around 15 Å, yielding quantum-efficiency variations over 2 orders of magnitude in the probed parameter region. Within the measured range of photon energies, $1.75 < h\nu < 4$ eV, there is a steep increase in the quantum efficiency between 2 and 2.5 eV and a comparably flat region above 2.5 eV, as shown in Fig. 5.8. These results agree only qualitatively with data obtained by Mort et al. [253]. It appears likely that the discrepancies are due to different oxygen concentrations in the films. Our results indicate that the states accessed with photon energies above ~2.2 eV act as effective transport states, allowing the photogenerated pair to escape from the Coulomb potential and contribute to the electric current.

5.3.2 The Nature of Transport States in C$_{60}$

Initial experiments to determine the band-gap in C$_{60}$ were carried out by Lof et al. [254] using photoemission and inverse photoemission measurements. Since the two experiments did not probe the same excitation state of the molecule, a somewhat complicated estimate for the excitonic binding energy

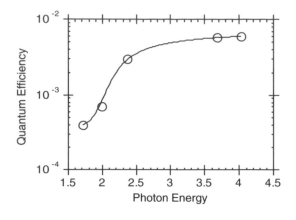

Fig. 5.8. Quantum generation efficiency versus photon energy in C$_{60}$ (Data from [250])

had to be employed to derive a band-gap of 2.3±0.1 eV from the experimental data. Since then, several investigations have been dedicated to the subject, and it is now believed that the transitions of 2.3 eV relate to charge-transfer states, i.e., excitonic states with the electron–hole separation covering more than a single molecule. The single-carrier band-gap may be as large as ∼2.6 eV [255,256]. Since these charge-transfer states involve several neighboring molecules they only arise in solids, and one may hope to detect these states in a comparison between molecular and solid-state absorption spectra. Figure 5.9 shows such a comparison involving molecular C$_{60}$ in benzene solution and thin solid films. At $T = 4.2$ K the thin-film absorption spectrum is sufficiently structured to allow a direct comparison to the solution spectrum. Both spectra are very similar near the absorption onset except for a small shift in energy, which can be attributed to the different dielectric constants. A 10 meV shift of the solution spectrum brings the two spectra to nearly perfect coincidence, as shown in Fig. 5.9b. The close similarity clearly indicates that the lower excited states are largely unaffected by solid-state effects, i.e. that they are excitonic molecular states. For photon energies in excess of ∼2.2 eV, there is strong absorption gain in the film. This absorption gain is interpreted as being due to the wavefunction overlap in charge transfer states [255]. Considering the nature of these states, we find the threshold value of 2.2 eV found here compatible with the assignment of 2.3 eV in [255].

A similar comparison at room-temperature gives essentially the same result [257]; there is therefore no discernible difference in the energy of the charge-transfer states in the low-temperature simple cubic phase and the high-temperature fcc phase. Figure 5.9a shows that the photocurrent spectrum also exhibits the vibronic structure as seen in the absorption spectra in the range below ∼2.2 eV, clearly indicating that the photoconductance in this regime is due to excitonic dissociation.

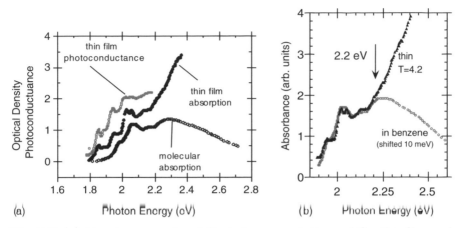

Fig. 5.9. (a) Absorbance spectra of C_{60} in benzene solution and C_{60} thin film, and the photoconductance spectrum of C_{60} thin film. (b) Solution spectrum shifted by +10 meV, and the thin-film absorbance spectrum. Note that scaling on the vertical axis corresponds to adjusting the differing optical thickness of the two samples. (Data from [232])

5.3.3 Mechanism of Charge Transport

The next question addresses the kinetics of carrier propagation. In Fig. 5.10 typical decay curves of the photoconductance obtained after pulsed laser excitation are shown. In the ns and ms time region, the photocurrent decay follows an algebraic decay law, indicative of dispersive transport conditions. The same decay kinetics are observed for all excitation energies in the range $2 < h\nu < 4$ eV and throughout the temperature range $100 < T < 300$ K. The absence of a strong T-dependence clearly indicates that thermal emission is not the rate-limiting process for carrier transport. This implies that multiple-trapping can be ruled out as a propagation mode and suggests hopping-type transport as a mechanism for the inter-molecular charge transfer. In the hopping transport model the algebraic decay results from the spatial distribution of sites and the strong dependence of the transfer integral on the distance between the sites. The hopping model therefore predicts the same kinetic behavior at all temperatures, in agreement with the data of Fig. 5.10.

It is interesting to note that the steady-state dark conductance and photoconductance also shows features typically associated with hopping transport. Figure 5.11 shows the results of steady-state measurements plotted versus $T^{-0.25}$, which is the T-dependence predicted for variable-range hopping [28]. A comparison with Fig. 5.1 shows that the experimental data fit the hopping-law much better than the Arrhenius behavior.

By considering the results stated in the previous paragraphs, we are led to the following picture of the charge transport in C_{60} films: Photoexcitation with energies around 2 eV gives rise to excitonic electron–hole pairs with a

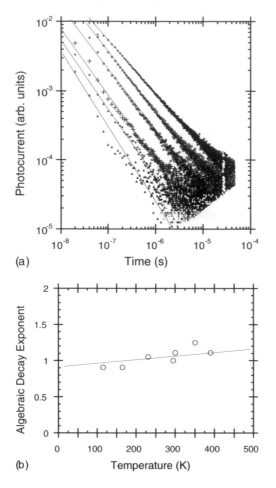

(a)

(b)

Fig. 5.10. (a) Photoconductance decay in C$_{60}$ films, following an algebraic decay law given by $i(t) \sim t^{-1}$. (b) T-dependence of the decay exponent indicates only a small T-dependence

binding energy of the order of $\sim$0.5 eV. This large value results in a low probability for thermal dissociation and low quantum generation efficiencies. The dissociation behavior is well described by the Onsager model. The generation efficiency strongly depends on the photon energy and the electric field, but extrinsic effects dominate for low fields. The carrier decay kinetics is the same for excitation energies in the range $2 < h\nu < 4$ eV, indicating that trapping and transport involve the same group of states. The T-dependence suggests that the transport occurs by hopping in localized states.

5.3.4 Electron and Hole Drift Mobilities and Recombination Times

For quantitative measurements of the transport parameters, the moving-photocarrier-grating technique has been used, which gives results for the

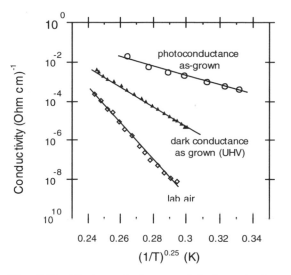

Fig. 5.11. Photoconductance and dark conductance data of Fig. 5.1 plotted versus $T^{-1/4}$ to confirm hopping propagation

electron and hole drift mobilities and the recombination time [88,110,175]. As in the case of a-Si:H, the experimental conditions were adjusted such that dielectric relaxation was slower than recombination. Figure 5.12 shows the experimental results obtained for a polymer-sealed film at different stages of O exposure after deposition. The solid lines show calculated fits based on a numerical evaluation of the transport and Poisson equations and a linear recombination ansatz [88,110]. A strong degradation of the transport parameters due to the influence of oxygen is noticeable. The initial electron drift mobility, measured within 20 min after deposition, is $\mu_e = 1.3$ cm^2(Vs)$^{-1}$, a value comparable to that of other polycrystalline organic semiconductor films [258,259] and compatible with the suggested transport by hopping. The hole mobility is found to be much smaller than the electron mobility. We find $\mu_h \approx 2 \times 10^{-4}$ cm^2(Vs)$^{-1}$, confirming that the hole contribution to dark conductance and photoconductance is very small [260].

The recombination time in the first measurement is 1.7 μs. This value is considerably larger than typical decay times from time-resolved photoconductance studies [261–263]. However, most of the kinetic measurements have focused on the ps and ns time region, and their basic result is that the decay becomes slower with progressing time. The data in Fig. 5.10 suggest an algebraic decay law, covering all of the ns time region and reaching several μs. Others have observed a stretched-exponential or multi-exponential decay behavior. A possible explanation for these observations is that fast trapping precedes a comparably slow recombination process. This explanation would

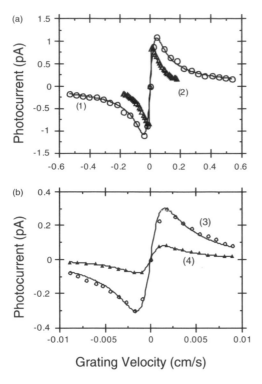

Fig. 5.12. Moving grating results for C$_{60}$ [88]. (**a**) Curve *1* shows results for a film $\sim$20 min after deposition ($\mu_c = 1.3$ cm^2(Vs)$^{-1}$, $\mu_h = 2 \times 10^{-4}$ cm^2(Vs)$^{-1}$, $\tau = 1.7$ μs), curve *2* shows results obtained after 5 h ($\mu_e = 0.45$ cm^2(Vs)$^{-1}$, $\mu_h = 7 \times 10^{-5}$ cm^2(Vs)$^{-1}$, $\tau = 5$ ms). (**b**) Results for more heavily degraded samples: curve *3* shows results obtained after 76 h; $\mu_e = 7 \times 10^{-5}$ cm^2(Vs)$^{-1}$, $\eta\mu\tau = 5 \times 10^{-10}$ cm^2 V^{-1}; curve *4* shows results obtained after 100 h; $\mu_e = 6 \times 10^{-5}$ cm^2(Vs)$^{-1}$, $\eta\mu\tau = 2 \times 10^{-10}$ cm^2 V^{-1}

also be in line with the finding of small $\mu\tau$ products under trapping conditions, as shown in Fig. 5.6 [250].

The product of quantum efficiency, drift mobility and recombination time, $\eta\mu\tau$, is a relevant photoconductance parameter for many applications, such as xerography, energy conversion or detector applications. For recombination conditions, $\eta\mu\tau \approx 7 \times 10^{-9}$ cm^2 V^{-1} is obtained in oxygen-free samples at low applied fields. $\eta\mu\tau$ is thus about a factor of 10 smaller than typical values in intrinsic a-Si:H.

5.4 Effect of Oxygen

Strong effects of oxygen on the photophysical properties of C$_{60}$ have been documented from the beginning of C$_{60}$ investigations [264–279]. Using the moving-grating technique in combination with quantum-efficiency measurements it has been possible to measure the oxygen-induced changes in the transport parameters [173,174,280]. These changes naturally depend on the rate at which oxygen is admitted to the samples, and this in turn depends on the sample structure. Two different sample configurations were used in the present work. The time-resolved experiments employed a sandwich structure with the C_{60} film between the conductive glass substrate and a $\sim$160 Å thick metal film, which was deposited without breaking UHV conditions. These films show stable electronic properties for a few days or more.

The moving-grating experiments employed a coplanar contact arrangement consisting of two parallel contact strips deposited on the glass substrate and contacted through holes in the C_{60} film. Several attempts were made to use SiO_2 and MgF as insulating cover layers on top of this structure to prevent in-diffusion of O. But is was found that these diffusion-barriers also affect the conductance, probably due to band-bending at the interface. Covering the samples with adhesive polymer tape in the deposition chamber was found to leave the conductance nearly unaffected. The moving-grating measurements were therefore performed on tape sealed samples. The photo conductance and quantum efficiency changes for the tape-sealed samples are shown in Fig. 5.13. Significant degradation in the photoconductance is seen to set in after approximately 5 h, as indicated by a sharp turn-over on the log-log plot of Fig. 5.13. The 5 h time bracket is, however, sufficient to obtain reproducible results on the transport parameters. Comparison among several measurements in the early time regime further indicates that the initial degradation is small.

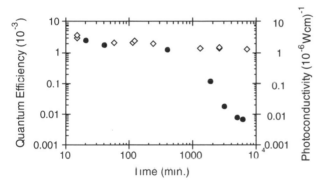

Fig. 5.13. Change of quantum generation efficiency and photoconductance in tape-sealed C_{60} films with exposure to air [88], ($\bullet$) Photoconductance, ($\Diamond$) quantum efficiency

Electron spin resonance (ESR) and secondary ion mass spectroscopy (SIMS) were used to characterize the O content in the as-grown and degraded state. The ESR signal results from a resonance with a g value of 2.0028 originating from the oxidized fullerene molecule, C_{60}^{+} [281–283]. Assuming the oxidation to be due to a single O atom, the ESR signal strength can be evaluated in terms of the chemically active O content in the films. In the as-grown state typical spin densities are in the 10^{15} cm^{-3} range, while the most strongly degraded samples have an ESR spin density of 10^{17} cm^{-3}.

Comparing the ESR results with concentration measurements by SIMS indicates that only 1 to 10% of the incorporated oxygen induces oxidation reactions.

The moving-grating technique is sufficiently sensitive to determine the transport parameters even for severely oxidized samples. Curve 2 in Fig. 5.12 shows results obtained for a lightly oxidized sample obtained after ~5 h in ambient atmosphere, and Fig. 5.12b more heavily degraded samples. A summary of the moving-grating results is shown in Fig. 5.14. The O-induced degradation involves a dramatic decrease in the carrier drift mobilities and an increase in the recombination time. It appears likely that these changes are linked to the charge transfer accompanying the oxidation reaction. Chemical studies have shown that the oxygen impurity state lies 0.3 eV below the excited triplet, $^3F_{1g}$ [284–287]. It therefore constitutes a deep defect state with an energy not far from midgap. This certainly explains the strong electronic compensation effect of O, apparent in the drastic lowering of the dark conductivity (Fig. 5.1). The charge transfer in the oxidized molecule gives rise to a negative net charge on the impurity, which may further affect the transport properties. As discussed for TiO$_2$, when localized charge concentrations reach values above ~10^{17} cm^{-3}, they may give rise to potential fluctuations of some 100 meV. Presently it is not known how such random potentials affect the carrier kinetics when the carriers propagate by hopping. It is to be expected, however, that in the case of transport-limited recombination kinetics, any decrease in mobility is linked to an increased recombination time, as observed in Fig. 5.14. Unlike the findings for TiO$_2$, the changes in recombi-

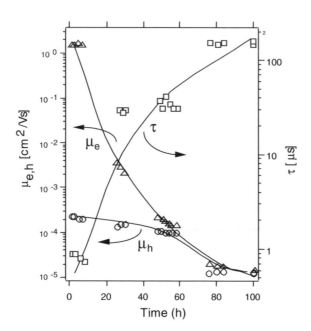

Fig. 5.14. Change in transport parameters due to O uptake in C$_{60}$ films. In the course of these measurements the ESR spin density was found to rise from ~10^{15} cm^{-3} to ~10^{17}cm^{-3} [88,173]

nation time are weaker than the mobility changes, such that the $\mu\tau$ product decreases with O uptake. These differences could, however, also be the result of other effects, such as changes in the charge-transfer rates in the vicinity of the impurity sites, as suggested in studies on photopolymerization and dimer formation in fullerenes [288,289].

5.5 Discussion of Transport Results

The results presented suggest that localized excitonic states in C_{60} play a dominant role in charge-carrier generation and transport. The photoconductance threshold involves the transition from the molecular ground state to the singlet Γ_{1g} states. There is a finite probability for a vibrationless transition into this state at 1.85 eV and increasing probability as one or more vibrations are coupled to the electronic transition. These lowest excited states appear to be localized. They do not give rise to a sizable signal in inverse photoemission, and carrier transport among these localized states is characterized by low quantum efficiencies.

The lowest states related to solid-state effects are the charge transfer states which are accessed with photon energies of approximately 2.2 eV. Excitation into these states is associated with increased quantum efficiencies, but the overall quantum efficiencies remain low up to excitation energies of 4 eV, indicating that all these states have excitonic character. This is nicely corroborated by the finding that the generation efficiency agrees with the Onsager model.

Measured electron mobilities are in the $1 \text{ cm}^2 (\text{Vs})^{-1}$ range and approximately two orders of magnitude larger than hole mobilities. The electron conduction mechanism is hopping. The $\eta\mu\tau$ product for recombination conditions is in the 10^{-9} to 10^{-8} cm^2 V^{-1} range and hence approximately 1–2 orders of magnitude lower than the value for high quality a-Si:H.

Oxygen acts as a compensating agent on the n-type conductivity and is found to degrade the transport properties significantly.

Since the low quantum efficiency results from intrinsic properties of the solid, there appears to be little prospect for substantial improvement of the transport properties of pure films. A more promising avenue for technological development lies in a modification of the molecular subunit or the structure of the solid. These ideas have indeed been pursued and promising results have been established. It has been found that the quantum efficiency in the higher fullerene films are substantially better than in C_{60} films [265]. From the electronic standpoint, the higher fullerenes may therefore hold more promise than C_{60}, although their synthesis is more difficult than that of C_{60}. Improved generation efficiencies are also obtained in compounds and alloys. These findings have triggered the preparation of a solar cell consisting of a polymer–fullerene hetero-junction along similar lines as followed by us in the eta-cell concept. In this type of cell, C_{60} is used as a photosensitizer to a conducting polymeric host, and a large-area interface is prepared on the molecular scale. The main ideas underlying this concept are presented next.

5.6 The Polymer–Fullerene Solar Cell

Charge transfer between the fullerenes and aromatic conducting polymers is expected to be efficient, because both material systems have an extended π-electron system, which is two-dimensional in the fullerenes and one-dimensional in the polymers. Over the last few years different charge transfer mechanisms between fullerenes and many polymers have been documented [290–302]. Some of these hold promise for specific photoelectric applications. Depending on the position of energy levels in the polymer, C_{60} can act as a dopant or photosensitizer [303]. When acting as a dopant, C_{60} is usually an acceptor due to its high electron affinity. The C_{60} conduction band then lies below the polymer valence band and charge transfer occurs from the ground state of the polymer. This charge transfer usually results in an increase of the polymer dark conductivity, while only minor changes in the photoconductance behavior are expected.

Photosensitization of the host material can be achieved by carrier transfer through excited states [295]. In particular, if the alignment of levels between host and C_{60} is such that one carrier type is preferably transferred, one has achieved a charge-separating molecular unit which may considerably enhance the dissociation rate for excitons. This latter mechanism is thought to be operative when the polymer MEH–PPV is doped by small amounts of C_{60}. In MEH–PPV, doping concentrations of $\sim$1% are sufficient to increase the photoconductivity by approximately an order of magnitude [304].

First devices exploiting these possibilities have been fabricated [5,304,305]. A planar structure, SnO_2/MEH-PPV/C_{60}/Au, which simply connects the dopant and the host to collecting contacts, has shown an energy-conversion efficiency of $\sim$0.04% [304]. Subsequent work, which involved a modification of the device geometry into an interpenetrating, phase-separated interface, gave a conversion efficiency close to 1% [5].

6. Conclusions

The electric transport properties of a-Si:H, porous TiO_2 and C_{60} have been studied by use of photoelectric methods. The materials have in common a large density of localized states, and the electronic transport is therefore slow and short-ranged. At present the properties of porous TiO_2 and C_{60} are clearly inferior to those of a-Si:H. Nonetheless, it appears that there is a range of applications in which these materials could be employed. This view rests on the finding that novel device structures can alleviate some of the transport problems. Reduced transport path lengths, carrier collection using majority-carrier transport, and extended internal field regions are the main guidelines for the fabrication of such devices.

In a-Si:H the transport parameters were summarized for optimized, device-grade material. The work on solar cells allowed a complete characterization of the electron and hole transport parameters, a determination of electric field profiles, and estimates for the deep- and shallow-trap capture-cross-sections. The experimental results suggest the following qualitative transport picture for a-Si:H: large densities of shallow states result in dispersive transport conditions for both electrons and holes, although in excellent material electron transport can also be non-dispersive at room temperature. For the description of transient phenomena it is useful to introduce a time-dependent drift mobility, but this concept is not easily applicable under steady-state conditions or for the description of devices. For these purposes it is necessary to separate between localized and mobile carriers. An accurate description of the complex carrier kinetics in devices can only be obtained on the basis of numerical techniques.

A more quantitative evaluation of the results shows that in undoped device-grade material drift mobilities are in the 0.5–2 $cm^2(Vs)^{-1}$ range for electrons and in the 0.005–0.05 $cm^2(Vs)^{-1}$ range for holes, depending somewhat on field strength and film thickness. These values are consistent with free-carrier mobilities of ~15 and ~1 $cm^2(Vs)^{-1}$ for electrons and holes, respectively. The carrier ranges are limited by deep localized states with a density of 10^{15}–10^{16} cm^{-3}. These states act as traps or recombination centers, depending on occupation. Under trapping conditions, the mobility-lifetime-products are of the order of 10^{-8} $cm^2 V^{-1}$. Electron $\mu\tau$-products are a typically a factor of 3–10 larger than the hole $\mu\tau$ products. Under recombination conditions, electron $\mu\tau$ products are in the 10^{-6} cm^2(Vs)$^{-1}$ range, while the hole $\mu\tau$ products are similar to those for trapping.

A comparison with non-optimized material indicates that the transport parameters can be significantly different from these optimized values. In un-doped material, electron and hole recombination $\mu\tau$ products scale inversely with the mid-gap density of states, indicating that mid-gap states are the dominant recombination centers. In doped a-Si:H, majority-carrier $\mu\tau$ products for recombination are enhanced, while minority-carrier $\mu\tau$ products are reduced. In alloys all transport properties, lifetimes and mobilities, are inferior to pure films. Wide-gap alloys exhibit charge storage effects on the time scale of days and can accommodate sufficiently large space charges for photoelectric memory applications. Structural changes induced by illumination or irradiation increase the deep defect density and reduce the carrier ranges, while the mobilities remain largely unaffected.

On the phenomenological level, TiO$_2$ and C$_{60}$ exhibit surprisingly similar properties. Both materials appear to have large densities of localized states. As in a-Si:H the transport is dispersive and the carrier kinetics strongly depend on the gap state occupation.

For porous TiO$_2$, the localized states appear to be due to lattice distortions on the grain surfaces and interfaces. Direct evidence for these comes from high-resolution electron microscopy. The finding of a broadened absorption edge and the evidence for dispersive transport support this finding. The electron drift mobilities are in the 10^{-4} to 10^{-7} cm^2(Vs)$^{-1}$ range, and recombination $\mu\tau$ products are around 10^{-10} cm^2 V^{-1}. Our data show that the drift mobility increases when the carrier density is raised, while the recombination time decreases, keeping the $\mu\tau$ product constant. This behavior can be explained in a simple transport model based on trap saturation in band-tail states. An evaluation of the experimental results in terms of the model suggests the existence of a conduction band-tail with $E_0 = 62$ meV, a free electron mobility of $\sim$2 cm^2(Vs)$^{-1}$ and a free-carrier recombination time of $\sim$70 ps. Modeling the experimental data in terms of carrier-induced potential fluctuations also allowed a description of the experimental data, but gave unreasonable model parameters.

The experimental data indicate that the transport properties in TiO$_2$ are inferior to those of a-Si:H and strongly suggest that it is preferable to use porous TiO$_2$ in majority-carrier transport applications. Following this idea, photosensitive hetero-junctions were prepared in which the band alignment is such that only electrons can be transferred into the TiO$_2$. The principal suitability of these structures for charge separation and transport was demonstrated. Hetero-junctions to PbS quantum dots were prepared as optically opaque thin films. The PbS quantum dots showed band-gap widening up to $\sim$2 eV from the 0.41 eV PbS bulk value and were found to exhibit excellent charge-transfer efficiencies. Sensitization of TiO$_2$ was also demonstrated with CdTe and CdS deposited as thin films from liquid solutions, and initial results for a novel type of solar cell, the eta-cell, were presented.

The section on C_{60} gave a quantitative description of the carrier transport in this material.

The main obstacle for efficient carrier transport in C_{60} is the large binding energy of the lower excitonic states. The molecular character of these states gives rise to very low quantum efficiencies. Spectral photoconductance and optical results show the E_{00} transition energy to be 1.85 eV and indicate charge-transfer states to be at ~2.2 eV. The excitonic binding energy is likely to be larger than 0.5 eV, which immediately explains the low quantum efficiencies.

Our photoelectric data show that for electric fields in excess of 10^3 V cm^{-1} the quantum generation process is well described in terms of the Onsager model; for smaller fields, excitonic dissociation is caused by scattering at extrinsic defects and surfaces. For photon energies below 2.2 eV the quantum efficiencies are typically in the 10^{-3} range.

The quantum efficiency rises significantly at 2.2 eV, indicating that different states are accessed with these photon energies. It was suggested that these states are excitonic charge-transfer states.

The time-resolved and steady-state photoconductance experiments indicate that hopping is the dominant transport mechanism in C_{60}. Since the time-resolved measurements do not show evidence for extended-state transport, it can be concluded that capture into localized states is faster than ~1 ns, and that the trap densities are larger than 10^{17} cm^{-3}. Electron drift mobilities around 1 cm^2(Vs)$^{-1}$ and hole drift mobilities in the 10^{-3} cm^2(Vs)$^{-1}$ range were determined by the moving-grating method. These values are also compatible with the interpretation of hopping transport.

Transport and electron spin resonance results show that C_{60} films are rapidly oxidized. The transport properties degrade significantly in the oxidation process. The oxygen uptake of the films leads to a decrease of the hole and electron drift mobilities to values in the 10^{-4} cm^2(Vs)$^{-1}$ range, while the recombination time is increased to values around 100 µs. These findings clearly indicate that, similar to the case of porous TiO_2, the recombination rate is transport-limited. The transport results are compatible with a suggestion from electro-chemical work, that O introduces a charged trap state with energy below the excited triplet state.

The low quantum efficiency found in pure C_{60} films is not necessarily observed in composite films and C_{60}-based hetero-junctions. These findings are the starting point for efforts to prepare photovoltaic devices involving C_{60}-polymer hetero-junctions [5]. Similar to the eta-cell these devices use a large interface as a charge-separating junction.

Many of the experimental results reported here were obtained with time-resolved photoconductance techniques. These are conceptually simple, but they suffer from the difficulty of deriving steady-state parameters. Only recently have sensitive steady-state techniques become available for the study of low-mobility transport. We used the moving-photocarrier-grating technique for the determination of transport parameters in a-Si:H alloys and in C_{60}. The evaluation of the experimental results requires a numerical treatment of the underlying transport equations and is thus considerably more complicated than in the time-resolved techniques. However, as soon as device applications are envisioned, it appears inevitable that numerical techniques will be developed for the description of the carrier kinetics, and it is then natural to use the more complicated characterization methods. It is therefore believed that the newer methods will find growing acceptance in the future.

References

1. J. Pankove (ed.), Semiconductors and Semimetals, Vol. 21C, (Academic Press, Orlando, 1984)
2. R. A. Street, Hydrogenated Amorphous Silicon (Cambridge University Press, Cambridge, 1991)
3. J. Kanicki (ed.), Amorphous and Microcrystalline Semiconductor Devices (Artech House, Boston, 1991)
4. R. Könenkamp and P. Hoyer, Mater. Res. Soc. Symp. Proc. 426, 551 (1996)
5. G. Yu, J. Gao, J. C. Hummelen, F. Wudl, and A. J. Heeger, Science 270, 1789 (1995)
6. S. C. Moss and J. F. Graczyk, Phys. Rev. Lett. 23, 1167 (1969)
7. J. C. Phillips, J. Non-Cryst. Sol. 34, 153 (1979)
8. A. H. Mahan, B. P. Nelson, R. S. Crandall, and D. L. Williamson, IEEE Trans. Elect. Devices 36, 2859 (1989)
9. C. Tsang and R. A. Street, Phil. Mag. B 37, 601 (1978)
10. C. R. Wronski, Sol. Energy Mat. 1, 287 (1979)
11. M. H. Brodsky, M. Cardona, and J. J. Cuomo, Phys. Rev. B 16, 3556 (1977)
12. G. Lucovsky, R. J. Nemanich, J. C. Knights, Phys. Rev. B 19, 2064 (1979)
13. E. C. Freeman and W. Paul, Phys. Rev. B 18, 8 (1978)
14. A. Madan, S. R. Ovshinsky, and E. Benn, Phil. Mag. B 40 259 (1979)
15. W. E. Spear, and P. G. LeComber, Sol. St. Comm. 17, 1193 (1975)
16. R. C. Chittick, J. H. Alexander, and H. F. Sterling, J. Electrochem. Soc. 116, 77 (1969)
17. J. C. Knights, Phil. Mag. 34, 663 (1976)
18. J. C. Knights, Jpn. J. Appl. Phys. 18, 101 (1978)
19. B. A. Scott, M. H. Brodsky, D. C. Green, P. B. Kirby, R. M. Plecenik, and E. E. Simonyi, Appl. Phys. Lett. 37, 725 (1980)
20. R. E. Rocheleau, S. S. Hegedus, W. A. Buchanan, and S. C. Jackson, Appl. Phys. Lett. 51, 133 (1987)
21. D. Kaplan, in: The Physics of Hydrogenated Amorphous Silicon I, J. D. Joannopoulos and G. Lucovsky (eds.) (Springer-Verlag, Berlin, 1984)
22. J. Lannin, J. Non-Cryst. Sol. 97&98, 39 (1987)
23. R. Könenkamp and M. Taguchi, unpublished results (1995)
24. J. D. Cohen, J. P. Harbison, and K. W. Wecht, Phys. Rev. Lett. 48, 109 (1982)
25. W. B. Jackson and N. M. Amer, Phys. Rev. B 25, 5559 (1982)
26. R. Biswas and D. R. Hamann, Phys. Rev. B 36, 6434 (1987)
27. G. D. Cody, T. Tiedje, B. Abeles, B. Brooks, and Y. Goldstein, Phys. Rev. Lett. 47, 1480 (1981)
28. N. F. Mott and E. A. Davis, Electronic Processes in: Non-Crystalline Materials (Oxford University Press, Oxford, 1979)
29. P. G. LeComber and W. E. Spear, Phys. Rev. Lett. 25, 509 (1970)
30. K. Winer, I. Hirabayashi, and L. Ley, Phys. Rev. B 38, 7680 (1988)

31. See for example: B. Stafford, and E. Sabisky (eds.), Proc. Int. Conf. Stability of Amorphous Silicon Alloy Materials and Devices, Palo Alto 1987 (American Institute of Physics Conf. Proc. 157, New York, 1987)
32. R. Könenkamp and E. Wild, Phys. Rev. B 42, 5887 (1990)
33. X. Xu and S. Wagner, in: Amorphous and Microcrystalline Semiconductor Devices, Vol. II, J. Kanicki (ed.) (Artech House, Boston 1992), Chap. 3
34. A. Werner, M. Kunst, and R. Könenkamp, Phys. Rev. B 33, 8878 (1986)
35. G. H. Bauer and G. Bilger, in: Proc. 5th Int. Symp. Plasma Chemistry, B. Waldie ed. (Heriot Watt Univ., Edinburgh, 1980) Vol. 2, p. 638
36. T. Shimada, S. Matsubara, H. Itoh, and S. Muramatsu, Proc. 9th EC Photovoltaic Solar Energy Conf., Freiburg (Kluwer Academic, Dordrecht, 1989), p. 81
37. H. Gleskova, R. Könenkamp, S. Wagner, and D. S. Shen, IEEE Electron Device Lett. 17, 264 (1996)
38. R. A. Caruso, M. Giersig, F. Willig and M. Antonietti, Langmuir Lett. 14, 6333 (1988)
39. R. Könenkamp, P. Hoyer, and A. Wahi, J. Appl. Phys. 79, 7029 (1996)
40. E. Pelizetti, C. Minero, E. Praauro, and M. Vincenti, in: Proc. Int. Conf. Photochemical and Photoelectrochemical Conversion and Storage of Solar Energy 1992 (International Academic Publishers, Bejing, 1993), p. 217
41. C. Bechinger, E. Wirth, and P. Leiderer, Appl. Phys. Lett. 68, 2843 (1996)
42. C. G. Granquist, Appl. Phys. A 57, 19 (1993)
43. R. Könenkamp and P. Hoyer, Mater. Res. Soc. Symp. Proc. 382, 147 (1995)
44. P. Hoyer and R. Könenkamp, Appl. Phys. Lett. 66, 349 (1994)
45. Y. Wang and N. Herron, J. Phys. Chem. 91, 257 (1987)
46. Y. Wang and N. Herron, Chem. Phys. Lett. 200, 71 (1992)
47. M. Meyer, C. Walberg, K. Kurihara, and J. H. Fendler, J. Chem. Soc. Chem. Comm. 90 (1984)
48. Y. Wang and N. Herron, J. Phys. Chem. 95, 525 (1991)
49. A. Henglein, Ber. Bunsenges. Phys. Chem. 99, 903 (1995)
50. L. E. Brus, R. W. Siegel et al., J. Mater. Res. 4, 705 (1989)
51. H. Weller, Angew. Chem. Int. Edition 32, 41 (1993)
52. L. N. Lewis, Chem. Rev. 93, 2693 (1993)
53. A. Henglein, J. Phys. Chem. 97, 5457 (1993)
54. J. H. Fendler, Advances in Polymer Science 113 (Springer Verlag, Berlin 1994)
55. C. Weisbuch and B. Vinter, Quantum Semiconductor Structures (Academic Press, San Diego, 1991)
56. G. P. Crawford and S. Zumer (eds.), Liquid Crystals in Complex Topologies, (Taylor and Francis, London, 1995)
57. B. O'Regan, J. Moser, M. A. Anderson, and M. Grätzel, J. Phys. Chem 94, 8720 (1990)
58. D. W. Bahnemann, C. Kormann, and M. R. Hoffmann, J. Phys. Chem. 91, 3789 (1987)
59. K. Ernst, Diploma thesis, Freie Universität Berlin (1997)
60. J. Fehrenbacher, Diploma thesis, Universität Konstanz (1997)
61. P. Hoyer, Dissertation, Technische Universität Berlin (1993)
62. R. Vogel, P. Hoyer, and H. Weller, J. Phys. Chem. 97, 7328 (1993)
63. R. Vogel, P. Hoyer, and H. Weller, J. Phys. Chem. 98, 3183 (1994)
64. H. Kroto, J. R. Heath, S. C. O'Brian, R. F. Curl and R. E. Smalley, Nature 318, 162 (1985)
65. W. Krätschmer, K. Fostiropolos, and D. R. Huffman, Chem. Phys. Lett. 170, 167 (1990)

66. F. Diederich, R. Ettl, Y. Rubin, R. L. Whetten, R. Beck, M. Alvarez, S. Anz, D. Sensharma, F. Wudl, K. C. Khemani, S. Hino, and A. Koch, Science 252, 548 (1991)
67. S. Iijima, Nature 354, 56 (1991)
68. T. W. Ebesen and P. M. Ajavan, Nature 358, 222 (1992)
69. J. Cioslowski, J. Am. Chem. Soc. 113, 6698 (1991)
70. R. E. Smalley, Nav. Res. Rev. 43 (III), 3 (1991)
71. J. M. Hawkins, A. Meyer, T. A. Lewis, S. Loren, F. J. Hollander, Science 252, 312 (1991)
72. K. E. Geckeler, and A. Hirsch, J. Am. Chem. Soc. 115, 3850 (1993)
73. B. Bhushan, B. K. Gupta, G. W. Van Cleef, C. Capp, and J. V. Coe, Appl. Phys. Lett. 62, 3253 (1993)
74. S. H. Friedman, D. L. DeCamp, G. L. Kenyon, R. Sibesma, G. Sranov, and F. Wudl, J. Am. Chem. Soc. 115, 6506 (1993)
75. M. N. Regueiro and P. Monceau, Nature 355, 237 (1992)
76. E. Sandre and F. Cyrot-Lackmann, Sol. St. Comm. 90, 431 (1994)
77. N. S. Sariciftci, Prog. Quant. Elect. 19, 131 (1995)
78. A. T. Werner, J. Anders, H. J. Byrne, W. K. Maser, M. Kaiser, A. Mittelbach, and S. Roth, Appl. Phys., A, 157 (1993)
79. P. A. Heiney, J. E. Fisher, A. R. McGhie, W. J. Romanow, A. M. Denenstein, J. P. McCauley, and A. B. Smith, Phys. Rev. Lett. 66, 2911 (1991)
80. H. Ogata, Y. Marayuma, T. Inabe, Y. Achiba, S. Suzuki, K. Kikuchi, and I. Ikemoto, Mod. Phys. Lett., 7, 1173 (1993)
81. A. F. Hebard, M. J. Rosseinsky, R. C. Haddon, D. W. Murphy, S. H. Glarum, T. M. Palstra, A. P. Ramirez, and A. R. Kortan, Nature 350, 600 (1991)
82. D. M. Cox, S. Behal, M. Disko, S. Gorun, M. Greaney, C. S. Hsu, E. Kollin, J. Millar, J. Robbins, W. Robbins, R. Sherwood, and P. Tindall, J. Am. Chem. Soc. 113, 2940 (1993)
83. F. Wudl and J. D. Thomson, J. Phys. Chem. Solids 53, 1449 (1992)
84. G. Yu, K. Pakbaz, and A. J. Heeger, Appl. Phys. Lett. 64, 3422 (1994)
85. W. Krätschmer, L. D. Lamb, K. Fostiropoulos, and D. R. Huffman, Nature 347, 354 (1990)
86. B. Pietzak, J. Erxmeyer, T. Almeida Murphy, D. Nagengast, B. Mertesacker, and A. Weidinger, in: Physics and Chemistry of Fullerenes and Derivatives, H. Kuzmany (ed.) (World Scientific, Singapore, 1995), p. 467
87. R. Könenkamp, J. Erxmeyer, and A. Weidinger, Appl. Phys. Lett. 65, 758 (1994)
88. G. Priebe, B. Pietzak, and R. Könenkamp, Appl. Phys. Lett. 71, 2160 (1997)
89. P. W. Anderson, Phys. Rev. 109, 1492 (1958)
90. J. M. Zimann, Models of Disorder (University Press, Cambridge, 1979)
91. K. B. Efetov, Adv. in Phys. 32, 53 (1983)
92. B. I. Shklovski and A. L. Efros, Electronic Properties of Doped Semiconductors, Springer Series Solid State Science, Vol. 45 (Springer-Verlag, Berlin, 1984)
93. I. M. Lifshitz, S. A. Gredescul, and L. A. Pastur, Introduction to the Theory of Disordered Systems (J. Wiley, New York, 1988)
94. D. Vollhardt and P. Wölfle, in: Electronic Phase Transitions, W. Hanke and Y. U. Kopaev (eds.) (North Holland, Amsterdam, 1991)
95. E. Abrahams, P. W. Anderson, D. C. Licciardello, and T. V. Ramakrishnan, Phys. Rev. Lett 42, 673 (1979)
96. E. Abrahams, P. W. Anderson, and T. V. Ramakrishnan, Phil. Mag. B42, 827 (1980)
97. D. J. Thouless, Phys. Rev. Lett. 18, 1167 (1977)
98. G. A. Thomas, Phil. Mag. B 52, 479 (1985)

99. D. Monroe, Phys. Rev. Lett. 54, 146 (1985)
100. R. J. Bandaranyake, G. W. Wen, J. Y. Lin, H. X. Jiang, and C. M. Sorensen, Appl. Phys. Lett. 67, 831 (1995)
101. M. Beaudoin, A. J. DeVries, S. R. Johnson, H. Laman, and T. Tiedje, Appl. Phys. Lett. 70, 3540 (1997)
102. G. D. Cody, in: Semiconductor and Semimetals, J. I. Pankove (ed.) (Academic Press 1984), Vol 21B, chap. 2
103. M. Ueta (ed.), Excitonic Processes in Solids (Springer-Verlag, Berlin, 1986)
104. J. P. Farges (ed.), Organic Conductors (Marcel Dekker, Inc., New York, 1994)
105. Z. Vardeny and J. Tauc, in: Semiconductors Probed by Ultrafast Laser Spectroscopy, R. R. Alfano (ed.) (Academic Press, New York, 1984) Vol. 2, p. 23
106. R. H. Bube, Photoelectric Properties of Semiconductors (Cambridge University Press, Cambridge 1992)
107. N. S. Sariciftci, Primary Photoexcitations in Conjugated Polymers (World Scientific, Singapore, 1997)
108. A. K. Jonscher, Dielectric Relaxation in Solids (Chelsea Dielectrics Press, London, 1983)
109. U. Haken, M. Hundhausen, and L. Ley, Appl. Phys. Lett. 63, 3066 (1993)
110. U. Haken, M. Hundhausen, and L. Ley, Phys. Rev. B 51, 10579 (1995)
111. A. Hoffmann and K. Schuster, Solid State Elect. 7, 717 (1964)
112. H. Benda, A. Hoffmann, and E. Spenke, Solid State Elect. 8, 887 (1965)
113. R. Könenkamp, A. M. Hermann, and A. Madan, Appl. Phys. Lett. 46, 405 (1985)
114. R. Könenkamp, and R. Henninger, Appl. Phys. A 58, 87 (1994)
115. A. Goodman, J. Appl. Phys. 32, 2550 (1961)
116. A. R. Moore, J. Appl. Phys. 54, 222 (1983)
117. J. Dresner, D. J. Szostak, and B. Goldstein, Appl. Phys. Lett. 38, 998 (1981)
118. T. J. McMahon and R. Könenkamp, Proc. 16th IEEE Photovoltaic Specialists Conf. (1982) p. 1389
119. G. H. Döhler, Phys. Rev. B 19, 2083 (1979)
120. W. B. Jackson, Phys. Rev. B 38, 3595 (1988)
121. R. S. Crandall, Phys. Rev. B 43, 4057 (1990)
122. H. Overhof and P. Thomas, Electronic Transport in Hydrogenated Amorphous Semiconductors (Springer-Verlag, New York, 1989)
123. A. Yelon, B. Movaghar, and H. M. Branz, Phys. Rev. B 46, 12244 (1992)
124. R. A. Street, J. Kakalios, and M. Hack, Phys. Rev. B38, 5603 (1988)
125. R. A. Street, M. Hack, and W. B. Jackson, Phys. Rev. B 37, 4209 (1988)
126. R. Zallen, The Physics of Amorphous Solids (John Wiley & Sons, New York, 1983)
127. M. Stutzmann, Festkörperprobleme 28, 1 (1988)
128. S. T. Pantelides, Phys. Rev. B 58, 1344 (1987)
129. S. B. Zhang, W. B. Jackson and D. J. Chadi, Phys. Rev. Lett. 65, 2575 (1990)
130. W. B. Jackson and J. Kakalios, Phys. Rev. B 37, 1020 (1988)
131. M. Stutzmann, W. B. Jackson and C. C. Tsai, Appl. Phys. Lett. 45, 1075 (1984)
132. D. Redfield, Appl. Phys. Lett. 52, 492 (1988)
133. G. Müller, S. Kalbitzer, and H. Mannsperger, Appl. Phys. A 39, 243 (1986)
134. Z. E. Smith, S. S. Aljishi, D. Slobodin, V. Chu, S. Wagner, P. M. Lenahan, R. R. Arya, and M. S. Bennett, Phys. Rev. Lett. 57, 2450 (1986)
135. T. J. McMahon and J. P. Xi, Phys. Rev. B 34, 2475 (1986)
136. R. Könenkamp and T. Shimada, Appl. Phys. Comm. 12, 11 (1993)

137. A. Mourchid, R. Vanderhagen, D. Hulin, and P. M. Fauchet, Phys. Rev. B 42, 7667 (1990)
138. M. Wraback and J. Tauc, Phys. Rev. Lett. 69, 3682 (1992)
139. H. Scher and E. W. Montroll, Phys. Rev. B12, 2455 (1975)
140. T. Tiedje, J. M. Cebulka, D. L. Morel, and B. Abeles, Phys. Rev. Lett 46, 1425 (1981)
141. J. Orenstein and M. A. Kastner, Phys. Rev. Lett. 46, 1421 (1981)
142. J. M. Hvam and M. H. Brodsky, Phys. Rev. Lett. 46, 371 (1981)
143. D. Monroe, and M. A. Kastner, Phil. Mag. B 47, 605 (1983)
144. H. Michiel and G. J. Adriaenssens, Phil. Mag. B 5, 27 (1985)
145. R. A. Street, J. Zesch, and M. J. Thomson, Appl. Phys. Lett. 43, 672 (1983)
146. P. B. Kirby and W. Paul, Phys. Rev. B 29, 826 (1984)
147. R. Könenkamp, S. Muramatsu, S. Matsubara, and T. Shimada, Appl. Phys. Lett. 60, 1120 (1992)
148. W. E. Spear and C. S. Cloude, Phil. Mag. B 58, 467 (1988)
149. R. I. Devlen, J. Tauc, and E. A. Schiff, J. Non-Cryst. Sol. 14, 507 (1989)
150. E. A. Schiff, R. I. Devlen, H. T. Grahn, and J. Tauc, Appl. Phys. Lett. 54, 1911 (1989)
151. G. Juska, G. Jukonis, and J. Kocka, J. Non-Cryst. Sol. 114, 354 (1989)
152. R. Stachowitz, W. Fuhs, and K. Jahn, Phil. Mag. B 62, 5 (1962)
153. C. Nebel, R. A. Street, N. M. Johnson, and C. C. Tsai, Phys. Rev. B 46, 6803 (1992)
154. P. B. Kirby, D. W. MacLeod, and W. Paul, Phil. Mag. B 51, 389 (1985)
155. R. A. Street, M. J. Thompson, and N. M. Johnson, Phil. Mag. B 51, 1 (1985)
156. R. Könenkamp, S. Muramatsu, H. Itoh, S. Matsubara, and T. Shimada, Appl. Phys. Lett. 57, 478 (1990)
157. R. Könenkamp, S. Muramatsu, H. Itoh, S. Matsubara, and T. Shimada, Jpn. J. Appl. Phys. Lett. 29, 2155 (1990)
158. K. Hecht, Z. Physik 77, 235 (1932)
159. R. Könenkamp, Phys. Rev. B 36, 2938 (1987)
160. R. Könenkamp, Proc. 19th Int. Conf. Physics of Semiconductors, Warsaw (1988), Inst. of Physics, Polish Academy of Sciences, Warsaw, p. 1621
161. A. Morimoto, T. Miura, M. Kumeda, and T. Shimuzu, J. Appl. Phys. 53, 7299 (1982)
162. M. Stutzmann, Phil. Mag. B 60, 531 (1989)
163. R. A. Street, J. Kakalios, and T. M. Hayes, Phys. Rev. B 34, 3030 (1986)
164. Z. E. Smith and S. Wagner, Phys. Rev. Lett. 59, 688 (1987)
165. R. Könenkamp and S. M. Paasche, Appl. Phys. Lett 49, 268 (1986)
166. C. van Berkel, in: Amorphous and Microcrystalline Semiconductor Devices, Vol. II, J. Kanicki (ed.) (Artech House, Boston, 1992), chap. 8
167. M. J. Powell, IEEE Trans. Elect. Devices 36, 2753 (1989)
168. H. Gleskova, R. Könenkamp, E. Y. Ma, S. Shen and S. Wagner, Dig. Tech. Papers, 3rd Annual Display Manufacturing Technology Conference, San Jose, CA, (1996) p. 97
169. J. Kocka, C. E. Nebel, and C. D. Abel, Phil. Mag. B 63, 221 (1991)
170. E. A. Schiff, Phil. Mag. Lett. 55, 87 (1987)
171. J. G. Simmons and G. W. Taylor, Phys. Rev. B4, 502 (1971)
172. C. R. Wronski and R. E. Daniel, Phys. Rev. B 23, 794 (1981)
173. G. Priebe, Diploma thesis, Technische Universität Berlin (1996)
174. G. Priebe and R. Könenkamp, Proc. Int. Conf. Lasers, Charlston 1995 (STS Press, McLean, VA. 1996), p. 494
175. L. Yang, A. Catalano, R. R. Arya, M. S. Bennett, and I. Balberg, Mater. Res. Soc. Symp. Proc. 149, 563 (1989)

176. R. Könenkamp, J. Non-Cryst. Solids 77/78, 643 (1985)
177. D. L. Staebler and C. R. Wronski, Appl. Phys. Lett. 31, 292 (1977)
178. H. Dersch, J. Stuke, and J. Beichler, Appl. Phys. Lett. 38, 456 (1980)
179. J. R. Macdonald, J. Appl. Phys. 62, R51 (1987)
180. R. Kohlrausch, Ann. Phys. 12, 393 (1847)
181. M. Hack and M. Shur, J. Appl. Phys. 58, 997 (1985)
182. M. Hack and M. Shur, J. Appl. Phys. 54, 5858 (1983)
183. H. Lin, S. Kumon, H. Kuzpka, and T. Yoko, Thin Solid Films 315, 266 (1998)
184. K. Tennakone, S. W. M. S. Wickramanayake, P. Samarasekara, and C. A. N. Fernando, Phys. Stat. Sol. A 104, K 57 (1987)
185. A. Stashans, S. Lunell, R. Bergström, A. Hagfeldt, and S. E. Lindquist, Phys. Rev. B 53, 159 (1996)
186. G. Nuspel, K. Yoshizawa, and T. Yamabe, J. Mater. Chem. 7, 2529 (1997)
187. S. Whittingham and M. B. Dines, J. Electrochem. Soc. 124, 1387 (1977)
188. A. Hagfeldt, N. Vlachopoulos, and M. Grätzel, J. Electrochem. Soc. 141, L82 (1994)
189. S. Huang, L. Kavan, I. Exnar, and M. Grätzel, J. Electrochem. Soc. 142, 142 (1995)
190. A. Hagfeldt, N. Vlachopoulos, S. Gilbert, and M. Grätzel, Proc. SPIE 2255, 297 (1994)
191. R. Könenkamp and A. Wahi, Mater. Res. Soc. Symp. Proc. 431, 467 (1996)
192. J. Yahia, Phys. Rev. 139, 1711 (1963)
193. E. M. Logothetis and W. J. Kaiser, Sens Actuators 4, 333 (1983)
194. A. V. Chadwick, R. M. Geatrches and J. D. Wright, Phil. Mag. A 64, 999 (1991)
195. D. Duongh, J. J. Ramsden and M. Grätzel, J. Am. Chem. Soc. 104, 2977 (1982)
196. S. E. Lindquist and H. Vidarsson, J. Mol. Catalysis 38, 131 (1986)
197. X. K. Zhao and J. H. Fendler, J. Phys. Chem. 95, 3716 (1991)
198. L. Spanhel, A. Henglein, and H. Weller, Ber. Bunsenges. Phys. Chem. 91, 1359 (1987)
199. A. Wahi, R. Engelhardt, P. Hoyer, and R. Könenkamp, Proc. 11th European Conf. Photovoltaic Science and Technology, Montreux (1992), p. 71
200. G. Rothenburger, D. Fitzmaurice, and M. Grätzel, J. Phys. Chem. 96, 5983 (1992)
201. R. Könenkamp, A. Wahi, and P. Hoyer, J. Phys. Chem 97, 7328 (1993)
202. R. Könenkamp, A. Wahi, and P. Hoyer, Thin Solid Films 246, 13 (1994)
203. Haipeng Tang, Dissertation, Ecole Polytechnique Federale de Lausanne (1995)
204. A. Rose, Photoconductivity and Allied Problems (R. E. Krieger Pub., Huntington, NY 1978)
205. D. E. Theodorou and H. J. Queisser, Phys. Rev. Lett. 58, 1992 (1987)
206. J. Jäckle, Phil. Mag. B 41, 681 (1980)
207. H. Overhof, J. Non-Cryst. Sol. 66, 261 (1984)
208. P. E. de Jongh and D. Vanmaekelbergh, Phys. Rev. Lett. 77, 3427 (1996)
209. A. Zaban, A. Meier, and B. A. Gregg, J. Phys. Chem. B 101, 7985 (1997)
210. K. Schwarzburg and F. Willig, Appl. Phys. Lett. 58, 2520 (1991)
211. A. Zaban, A. Meier, and B. A. Gregg, J. Phys. Chem. B 101, 7985 (1997)
212. A. Solbrand, H. Lindström, H. Rensmo, A. Hagfeldt, and S. Lindquist, J. Phys. Chem. B 101, 2514 (1997)
213. J. Nelson, Phys. Rev. B 59, 15374 (1999)
214. K. D. Schierbaum, U. K. Kirner, J. F. Geiger, and W. Göpel, Sens. Actuators B 4, 87 (1991)

215. Y. Wang, A. Suna, W. Mahler, and R. Kasowski, J. Chem. Phys. 87, 7319 (1987)
216. R. V. Kasowski, M. H. Tsai, T. N. Ahodin, and D. D. Chambiss, Phys. Rev. B 34, 2656 (1986)
217. H. M. Schmidt and H. Weller, Chem. Phys. Lett. 129, 615 (1986)
218. D. Schoos, A. Mews, A. Eychmüller, and H. Weller, Phys. Rev. B 49, 17072 (1994)
219. M. Giersig, private communication
220. M. A. Green, Solar Cells (Prentice Hall, Englewood Cliffs, NJ, 1982)
221. S. Siebentritt, K. Ernst, C. H. Fischer, R. Könenkamp, and M. C. Lux- Steiner, in: Proc. 14 European Photovoltaic Science and Technology Conf., Barcelona, Spain (1997) (Information Press, Eynsham) p. 1823
222. B. O'Regan, and M. Grätzel, Nature 353, 737 (1991)
223. K. Tennakone, G. Kumara, A. R. Kumarasinghe, K. Wijayantha, and P. M. Sirimanne, Semicond. Sci. Technol. 10, 1689 (1995)
224. B. O'Regan and D. T. Schwartz, Chem. Mater 7, 1349 (1995)
225. C. Rost, I. Sieber, S. Siebentritt, M. C. Lux-Steiner, and R. Könenkamp, Appl. Phys. Lett. 75, 692 (1999)
226. J. Möller, C. H. Fischer, S. Siebentritt, R. Könenkamp, and M. C. Lux- Steiner, Proc. 2nd World Conf. and Exhibition on Photovoltaics, Vienna (1998) (European Commission Ispra) p. 209
227. S. R. Morrison, Electrochemistry at Semiconductor and Oxidized Metal Electrodes (Plenum Press, New York, 1980)
228. T. Arai, Y. Murakami, H. Suematsu, K. Kikuchi, Y. Achiba, and I. Ikemoto, Sol. St. Comm. 84, 827 (1992)
229. M. Kaiser, W. K. Maser, H. J. Byrne, A. Mittelbach, and S. Roth, Sol. St. Comm. 87, 281 (1993)
230. Y. Chen, A. Kortan, R. Haddon, and D. Fleming, J. Phys. Chem. 96, 1016 (1992)
231. J. Mort, R. Ziolo, M. Machonkin, D. R. Huffman, and M. I. Ferguson, Chem. Phys. Lett. 186, 284 (1991)
232. R. Könenkamp, J. Erxmeyer, and A. Weidinger, Mater. Res. Soc. Symp. Proc. 349, 357 (1994)
233. N. J. Turro, Modern Molecular Photochemistry (University Science Books, Mill Valley, 1991), p. 467
234. K. Prassides, J. Thomkinson, C. Christides, M. Rosseinsky, D. W. Murphy, and R. C. Haddon, Nature 354, 462 (1991)
235. K. Prassides, T. J. Dennis, J. P. Hare, J. Tomkinson, H. W. Kroto, R. Taylor, and D. Walton, Chem. Phys. Lett. 187, 455 (1991)
236. S. Saito and A. Oshiyama, Phys. Rev. Lett. 66, 2637 (1991)
237. R. Saito, G. Dresselhaus, and M. S. Dreselhaus, Phys. Rev. B 46, 9906 (1992)
238. K. Prassides, C. Christides, M. Rosseinsky, J. Tomkinson, D. W. Murphy, and R. C. Haddon, Europhys. Lett. 19, 629 (1992)
239. R. C. Haddon, L. E. Brus, and K. Raghavachari, Chem. Phys. Lett. 125, 459 (1986)
240. G. Herzberg and E. Teller, Z. Phys. Chem. 21, 410 (1933)
241. G. Orlandi and W. Siebrand, J. Chem. Phys. 58, 4513 (1973)
242. K. Yabana and G. F. Bertsch, Chem. Phys. Lett. 197, 32 (1992)
243. F. Negri, G. Orlandi, and F. Zerbetti. J. Chem. Phys. 97, 6496 (1992)
244. P. C. Eklund, P. Zhou, K. Wang, G. Dreselhaus, and M. S. Dresselhaus, J. Phys. Chem. 53, 1391 (1992)
245. H. Schaich, M. Muccini, J. Feldmann, H. Bässler, E. O. Göbel, R. Zamboni, C. Taliani, J. Erxmeyer, and A. Weidinger, Chem. Phys. Lett. 236, 135 (1995)

246. M. Muccini, R. Danieli, R. Zamboni, C. Taliani, H. Mohn, W. Müller, and H. U. ter Meer, Chem. Phys. Lett. 245, 107 (1995)

247. C. Reber, L. Yee, J. McKiernan, J. I. Zink, R. S. Williams, W. M. Tong, D. Ohlberg, R. L. Whetten, and F. Diederich, J. Phys. Chem 95, 2127 (1991)

248. Y. Wang, J. M. Holden, A. M. Rao, P. C. Eklund, U. D. Venkateswaran, D. Eastwood, R. Lidberg, G. Dresselhaus, and M. S. Dresselhaus, Phys. Rev. B 51, 4547 (1995)

249. S. G. Louie and E. L. Shirley, J. Phys. Chem. Sol. 54, 1767 (1993)

250. R. Könenkamp, R. Engelhardt, and R. Henninger, Sol. St. Comm. 97, 285 (1996)

251. L. Onsager, Phys. Rev. 54, 554 (1938)

252. D. M. Pai and R. C. Enck, Phys. Rev. B 11, 5163 (1975)

253. J. Mort, M. Machonkin, I. Chen and R. Ziolo, Phil. Mag. Lett. 67, 77 (1993)

254. R. W. Lof, M. A. van Veenendaal, B. Koopmans, H. T. Jonkman, and G. A. Sawatzky, Phys. Rev. Lett. 68, 3924 (1992)

255. X. Wei, D. Dick, S. A. Jeglinski, and Z. V. Vardeny, Synth. Met. 86, 2317 (1997))

256. D. Dick, X. Wei, S. Jeglinski, R. E. Brenner, Z. V. Vardeny, D. Moses, V. I. Sradanov, and F. Wudl, Phys. Rev. Lett. 73, 2760 (1994)

257. R. Könenkamp, unpublished results

258. N. Karl, in: Defect Control in Semiconductors, K. Sumino (ed.) (Elsevier Science Publishers, North Holland, 1990), p. 1725

259. W. Warta, R. Stehle, and N. Karl, Appl. Phys. A 36, 163 (1985)

260. D. Sarkar and N. J. Halas, Sol. St. Comm. 90, 261 (1994)

261. C. H. Lee, G. Yu, D. Moses, V. I. Srdanov, X. Wei, and Z. V. Vardeny, Phys. Rev. B 48, 8506 (1993)

262. R. A. Cheville and N. J. Halas, Phys. Rev. B 4548 (1992)

263. C. H. Lee, G. Yu, B. Kraabel, D. Moses, and V. I. Srdanov, Phys. Rev. B 49, 10572 (1994)

264. G. H. Kroll, P. Benning, Y. Chen, T. Ohno, J. Weaver, L. Chibante, and R. E. Smalley, Chem. Phys. Lett. 181, 112 (1991)

265. M. Hosoya, K. Ichimura, Z. H. Wang, G. Dresselhaus, and M. S. Dresselhaus, Phys. Rev. B 49, 4981 (1994)

266. A. Hamed, Y. Y. Sun, Y. K. Tao, R. L. Meng, and P. H. Hor, Phys. Rev. B 47, 10873 (1993)

267. N. Minami and M. Sato, Synth. Met., 55–57, 3092 (1993)

268. P. Zhou, A. M. Rao, K. -A. Wang, J. D. Robertson, C. Eloi, M. S. Meier, S. L. Ren, X. X. Bi, and P. C. Eklund, Appl. Phys. Lett. 60, 2871(1992)

269. M. Kaiser, W. K. Maser, H. J. Byrne, A. Mittelbach, and S. Roth, Sol. St. Comm. 87, 281 (1993)

270. S. V. Chekalin, Appl. Phys. Lett. 71, 1276 (1997)

271. E. A. Katz, V. Lyubin, D. Faiman, S. Shutina, A. Shames, and S. Goren, Sol. St. Comm. 100, 781 (1996)

272. L. Akselrod, H. J. Byrne, C. Thomson, A. Mittelbach, and S. Roth, Chem. Phys. Lett. 212, 384 (1993)

273. G. H. Kroll, P. J. Benning, Y. Chen, T. R. Ohno, J. H. Weaver, L. P. F. Chibante, and R. E. Smalley, Chem. Phys. Lett. 181, 112 (1991)

274. P. Kovarik, E. B. D. Bourdon, and R. H. Prince, Phys. Rev. B 49, 7744 (1994)

275. H. Werner, M. Wohlers, D. Bublak, J. Blöcker, R. Schlögl, Fullerene Sci. Tech. 1, 457 (1993)

276. A. M. Rao, K. -A. Wang, J. M. Holden, Y. Wang, P. Zhou, P. C. Eklund, C. C. Eloi, J. D. Robertson, J. Mater. Res. 8, 2277 (1993)

277. C. C. Eloi, J. D. Robertson, A. M. Rao, P. Zhou, K. -A. Wang, P. C. Eklund, J. Mater. Res. 8, 3085 (1993)
278. A. M. Rao, M. Menon, K. -A. Wang, P. C. Eklund, K. R. Subbaswamy, D. S. Cornett, M. A. Duncan, I. J. Amster, Chem. Phys. Lett. 224, 106 (1994)
279. V. Vijayakrishnan, A. K. Santra, T. Pradeep, R. Seshadri, R. Nagarajan, and C. N. R. Rao, J. Chem. Soc. Chem. Comm., 198 (1992)
280. R. Könenkamp, G. Priebe, and B. Pietzak, Phys. Rev. B 60, 11804 (1999)
281. M. Baumgarten, A. Gügel, and L. Ghergel, Adv. Mat. 5, 458 (1993)
282. S. Kawata, K. Yamauchi, S. Suzuki, K. Kikuchi, H. Shiromaru, M. Katada, K Saito, I. Ikemoto, and Y. Achiba, Chem. Lett. 1659 (1992)
283. S. Kukolich and D. Huffman, Chem. Phys. Lett. 182, 263 (1991)
284. J. W. Arbogast, A. P. Darmanyan, C. S. Foote, Y. Rubin, F. N. Diederich, M. M. Alvarez, S. J. Anz, and R. L. Whetten, J. Phys. Chem. 95, 11 (1991)
285. M. Terazima, N. Hirota, H. Shinohara, and Y. Saito, J. Phys. Chem. 95, 9080 (1991)
286. Y. Kajii, T. Nakagawa, S. Suzuki, Y. Achiba, K. Obi, and K. Shibuya, Chem. Phys. Lett. 281, 100 (1991)
287. M. R. Fraelich and R. B. Weisman, J. Phys. Chem. 97, 11145 (1993)
288. A. M. Rao, P. Zhou, K. A. Wang, G. T. Hager, J. M. Holden, Y. Wang, W. T. Lee, X. X. Bi, P. C. Eklund, D. S. Cornett, M. A. Duncan, and I. J. Amster, Science 259, 955 (1993)
289. R. Taylor, J. Parsons, A. Avent, S. Rannard, T. Dennis, J. P. Hare, H. W. Kroto, and D. Walton, Nature 351, 277 (1991)
290. Y. Wang, Nature 356, 585 (1992)
291. B. Kraabel, C. H. Lee, D. Moses, N. S. Sariciftci, and A. J. Heeger, Chem. Phys. Lett. 213, 389 (1993)
292. Y. Wang, R. West, and C. H. Yuan, J. Am. Chem. Soc. 115, 3844 (1993)
293. L. Smilowitz, N. S. Sariciftci, R. Wu, C. Gettinger, A. J. Heeger, and F. Wudl, Phys. Rev. B 47, 13835 (1993)
294. C. S. Kuo, F. G. Wakim, S. K. Sengupta, and S. K. Tripathy, Sol. St. Comm. 87, 115 (1993)
295. S. Morita, A. Zakhidov, and K. Yoshino, Sol. St. Comm. 82, 249 (1992)
296. Y. Yamashita, W. Takashima, and K. Kaneto, Jpn J. Appl. Phys. Lett. 32, 1017 (1993)
297. K. Yoshino, X. H. Yin, S. Morita, T. Kawai, and A. Zakhidov, Sol. St. Comm. 85, 85 (1993)
298. S. M. Silence, C. A. Walsh, J. C. Scott, and W. E. Moerner, Appl. Phys. Lett. 61, 2967 (1992)
299. R. G. Kepler and P. A. Cahill, Appl. Phys. Lett. 63, 1552 (1993)
300. C. H. Lee, G. Yu, D. Moses, and A. J. Heeger, Appl. Phys. Lett. 65, 664 (1994)
301. C. H. Lee, G. Yu, D. Moses, K. Pakpaz, C. Zhang, N. S. Sariciftci, A. J. Heeger, and F. Wudl, Phys. Rev. B 48, 15425 (1993)
302. N. S. Sariciftci, L. Smilowitz, A. J. Heeger, F. Wudl, Science 258, 1474 (1992)
303. A. A. Zakhidov, H. Araki, K. Tada, K. Yoshino, Synth. Met. 77, 127 (1996)
304. N. S. Sariciftci, D. Braun, C. Zhang, V. I. Srdanov, A. J. Heeger, G. Stucky, and F. Wudl, Appl. Phys. Lett. 62, 585 (1993)
305. C. S. Kuo, F. G. Wakim, S. K. Sengupta, and S. K. Tripathy, Sol. St. Comm. 87, 115 (1993)

Index

Springer Tracts in Modern Physics